The evolution of mechanics

Monographs and textbooks on mechanics of solids and fluids

editor-in-chief: G. Æ. Oravas

Mechanics: Genesis and method

editor: G. Æ. Oravas

1. P.-M.-M. DUHEM
   The evolution of mechanics

The evolution of mechanics

Pierre-Marie-Maurice Duhem
Professor of theoretical physics

Introduction by G. Æ. Oravas

translated by

Michael Cole

SIJTHOFF & NOORDHOFF   1980
Alphen aan den Rijn, The Netherlands
Germantown, Maryland, USA

Copyright © 1980 Sijthoff & Noordhoff International Publishers B.V., Alphen aan den Rijn, The Netherlands

All rights reserved. No part of this publication may be reproduced, stored in a retrieval system, or transmitted, in any form or by any means, electronic, mechanical, photocopying, recording or otherwise, without the prior permission of the copyright owner.

ISBN 90 286 0688 2

Original title:
"L'Evolution de la mécanique", published by Joanin, Paris in 1903; reissued by A. Hermann, Paris in 1905.

PHYSICS DEPT.
QA
802
.D8713

Printed in The Netherlands

CONTENTS

EDITOR'S INTRODUCTION... ix
PHOTOGRAPHS OF PIERRE-MARIE-MAURICE DUHEM. xxxiv
TRANSLATOR'S NOTE. xxxvii
AUTHOR'S INTRODUCTION... xl

PART ONE

MECHANICAL EXPLICATIONS

Chapter

I	PERIPATETIC MECHANICS.	1
II	CARTESIAN MECHANICS...	5
III	ATOMISTIC MECHANICS...	9
IV	NEWTONIAN MECHANICS...	12
V	FORCE AND HIDDEN PROPERTIES.	16
VI	THE PRINCIPLE OF VIRTUAL VELOCITIES AND LAGRANGE'S STATICS..	22
VII	D'ALEMBERT'S PRINCIPLE AND LAGRANGE'S DYNAMICS ...	32
VIII	LAGRANGE'S ANALYTICAL MECHANICS AND POISSON'S PHYSICAL MECHANICS..	37
IX	THE KINETIC THEORY OF GASES.	47
X	THE MECHANICAL THEORY OF HEAT	54
XI	THE MECHANICAL THEORIES OF ELECTRICITY..	68
XII	THE IMPOSSIBILITY OF PERPETUAL MOTION...	73
XIII	HERTZ'S MECHANICS	83
XIV	THE VORTEX ATOM.	89
XV	GENERAL CONSIDERATIONS ON MECHANICAL EXPLICATIONS...	94

PART TWO

THERMODYNAMICAL THEORIES

Chapter

I	THE PHYSICS OF QUALITY	105
II	ON THE COMPARISON OF THEORY WITH EXPERIMENT, AND ON VIRTUAL MODIFICATION	111
III	EQUILIBRIUM AND MOTION	116
IV	THE CONSERVATION OF ENERGY	118
V	WORK AND QUANTITY OF HEAT	122
VI	THE REVERSIBLE MODIFICATION	127
VII	CARNOT'S PRINCIPLE AND ABSOLUTE TEMPERATURE	131
VIII	INTERNAL POTENTIAL AND GENERAL STATICS	134
IX	THE PRINCIPLE OF GENERAL DYNAMICS	141
X	SUPPLEMENTARY RELATIONS	146
XI	THE KINETIC ENERGY EQUATION AND USABLE ENERGY	150
XII	STABILITY AND DISPLACEMENT FROM EQUILIBRIUM	153
XIII	FRICTION AND FALSE CHEMICAL EQUILIBRIA	161
XIV	PERMANENT ALTERATIONS AND HYSTERESIS	170
XV	ELECTRODYNAMICS AND ELECTROMAGNETISM	178
CONCLUSION		185
SUBJECT INDEX		191

Pierre Duhem was not only a scientist, philosopher and historian, he was also a talented cartoonist. As a homage paid by one cartoonist to another, this is the editor's cartoon of Pierre Duhem, when Duhem was a university student and famous as a caricaturist.

EDITOR'S INTRODUCTION

PIERRE DUHEM: SCIENTIST — PHILOSOPHER — HISTORIAN

In 1903 a long series of seven articles appeared by the pen of Pierre-Marie-Maurice Duhem (1861-1916) in Volume 14 of the journal *Revue générale des sciences*. They were written at the invitation of L. Olivier, the editor of that journal, as an effort to clarify the uncertain state of rational mechanics at that time. These articles were immediately published as a book entitled *L'évolution de la mécanique* by the publishing house Joanin in Paris in 1903, and subsequently reissued in 1905 by the publishing house A. Hermann in Paris. This important historico-critical exposition was originally intended for the edification of the professional men of the time, but it has retained its great value as a brilliant historico-critical introduction to the mechanical explanation of physical sciences up to this very day. It was written by one of the most gifted teachers of physical and chemical sciences of any period, a rational mechanician who was also an incredibly productive leading scientist, the first modern historian of science, and the leading philosopher of science. As a combination of all these four professional qualities in one person, Duhem has had no equals since antiquity. He was, at the same time, a very controversial figure in the French scientific community, and until very recently little had been written in the English language about this seminal thinker. His talents were so unusual, and his professional life so unconventional, that it is necessary to give a brief sketch of his life and a brief overview of his prodigious professional efforts in order to understand the book that is offered here in an English translation.

Pierre-Marie-Maurice Duhem, like Josiah Willard Gibbs (1839-1903), the American physicist and mathematician whose work inspired Duhem's own researches, was a man of extraordinary genius, of a kind not seen since his death from a heart attack in 1916. His biographers have been puzzled about his character, and some of them have called him a right-winger, a royalist, an anti-Republican, a nationalist, an anti-Semite, and a religious extremist. These supposed attitudes of Duhem have been explained as blind prejudices which he inherited from his bourgeois family. Such views, even expressed by biographers who were quite sympathetic to Duhem, stem from a certain lack of a deeper understanding of the kind of strong Christian beliefs that motivated Duhem's actions, actions which on occasion seemed dramatic and drastic to people who lacked Duhem's profound Christian convictions. Without this Christian viewpoint, the behaviour of Duhem on several occasions tends to contradict his supposedly blind prejudices. This political pidgeon-holing of Duhem, a man of profound Christian convictions and an extraordinarily wide and deep Western cultural perspective, is quite unsatisfactory, because the reverend attitude of such a well-bred individual as Duhem towards his beloved Western civilisation could have scarcely been motivated by such simple blind emotions as this rather facile assessment of his character implies. Duhem as a person can only be understood when it is realised that he was a true believer in mediaeval Christianity. As an academic he represented a most unusual figure, for he was as fiercely independent and as proudly individualistic as the Vikings who conquered the Franks, and he had the moral principles and manners of a Christian gentleman, and the valour of a Christian knight who was ready to crusade at a moment's notice when his Christian principles were challenged. Duhem had a high regard for the accomplishments of the Middle Ages, a period he had thoroughly studied, and he knew all its intimate details well. He readily recognised that Western civilisation was a Christian civilisation which was culturally fused with Hellenism, and as a civilisation it was a creation of the Middle Ages — the Catholic Middle Ages — a fact that strongly impressed him. Duhem easily recognised that the early part of the Middle Ages was the most precocious period in the history of man, a period of unprecedented spiritual, technological, and scientific advance in the life of man. He was quite aware that one of the most important ideas, the Christian concept of the independend private individual with an inviolable private conscience — a private inner world of the individual man, an inner sanctum, in which resides his personal worth and his dignity as a man — was born from the Christian belief of the Middle Ages. The spiritual independence of the individual man, the spiritual equality and value of every individual man as a child of God, and the freedom of choice of man — as an individual — about the conduct of his personal life, were the basic Christian beliefs which Duhem learned to appreciate from his studies of Western civilisation, and he held them in the highest regard. All his life, Duhem recognised the great importance of the Christian principle of charity as a voluntary personal act of being merciful and forgiving to the penitent, and avoiding hypocrisy as a deadly sin. He passionately believed in the Christian responsibility of the individual man to seek the truth, for Christ had said that only the truth can make a man free, to stand up for justice regardless of the personal costs, and to uphold his personal Christian traditions without exception. However, to a sound Christian such as Duhem, to forgive and not to forget was a *sine qua non*.

Duhem was fully aware that Christianity was by far the most important part of Western culture, for it is belief in Christian moral principles that essentially makes Western Man. To a true believer such as Duhem, anything that ran counter to these basic Christian beliefs of Western Man eroded the Christian tradition of Western society, and, therefore, undermined Western civilisation as a Christian civilisation. These Christian views according to which Duhem acted were of mediaeval origin, and he was absolutely fearless in upholding and fighting for them in his professional life. In this regard Duhem appears as a figure from the remote past: a veritable reincarnation of the Frankish Christian knight on a crusade, for whenever his fundamental Christian principles were at stake he was quite willing to go even as far as physical violence to uphold them. On any point of principle, Duhem was absolutely unyielding and very unlike his academic colleagues, most of whom seemed to have been infinitely compliant to the will of the institutional bureaucracy and the whims of important state functionaries. Duhem, on the other hand, seems to have put on the mantle of his Christian responsibilities in opposing with unflinching conviction any anti-Christian tendencies, whenever they arose and wherever they were found. For instance, Duhem refused to help George Sarton in the launching of *Isis*, a liberal journal of science history, because of his religious principles. Duhem certainly considered the liberal free-thinkers of his time as the ideological offspring of the totalitarian Jacobins of the First Republic, a totalitarian democracy, and to him they were the most prominent promoters of paganism in French society since the time of the Enlightenment. He seems to have held the same view of atheistic Marxists. Perhaps this Christian view explains Duhem's anti-liberalism the best. As the French Republic represented a tradition of totalitarian democracy, supported by universal suffrage, which refused to recognise the God-given rights of the Christian individual, and which violated the Christian decalogue by deliberately promoting among the common people vulgar greed and unchristian envy of the decent productive citizen, it must have been one of the main reasons that motivated his anti-Republicanism. His opposition to universal suffrage could have been influenced by his critical examination of the totalitarian democracy which had undermined the Christian principles and dignity of the common Frenchman in the street, and had made him dishonest and greedy; as a consequence of his paganisation, the common man had then tended to use the right to vote as a licence to loot the decent, productive Christian of the fruits of his labour. That Duhem considered this plunder to be sinful and an unqualified evil is quite obvious; this view and the fact that universal suffrage supported such an anti-Christian political system may have been enough for Duhem to oppose it.

He probably opposed the Christian-Democratic movement because to him it must have represented an inner logical contradiction: a political system such as a totalitarian democracy which violates fundamental Christian principles cannot be Christian. Perhaps Duhem considered Judaism as an organised religious movement within the gates of Christian society to be inimical to the well-being of the Christian society, because Judaism condemns Jesus as an imposter, denies Jesus' divinity, and therefore regards Christianity as a fraudulent religion. The tendency of this organised religious opinion existing within Christian society is to create doubt amongst the Christians about their Christian convictions. This consideration — which has mediaeval roots — most likely accounts for what is called Duhem's anti-semitism. Duhem was also unabashedly nationalistic; in this attitude, in contrast to his Christian attitude, he seems almost to have been *like* a contemporary academic. In the popular writings of Duhem which contained war-time propaganda — such as the small brochure, *La chimie est-elle une science française?*, in which he proudly contested the claim of the German physical chemist Wilhelm Ostwald (1853-1932) that chemistry was a German science, and in the small book, *La science allemande*, in which he criticises obscurantism and "Naturphilosophie" in German science — he never resorted to vicious language or invective, as common then as today. Duhem remained a convinced French nationalist all his life, and it has to be admitted that his fervent patriotism definitely influenced his scientific opinions — but never to the extent of making them totally invalid. Despite these firm Christian attitudes, Duhem was a highly civilised Western man who was not intolerant of other people with views differing from his own, just as long as they did not take overt action to undermine his beloved Christian civilisation by aggressing against the dignity of a Christian person or by trying to refute the Christian principles exemplified by Mediaeval Catholicism, which were to Duhem the very underpinning of Western Christian civilisation. This deeply felt Christian morality of Duhem does explain why he could feel so friendly and be so helpful to people whose principles and backgrounds were such that Duhem should have disliked them on sight. Whenever Duhem saw the principles of personal dignity and justice being violated he went to almost any length to defend them. Since the dogmatic Duhem did not believe in the misconstrued 'turn-the-other-cheek' Christianity, he did not allow any misrepresentation of the truth, any deed of injustice, or any cynical attempt to violate the dignity of man, to pass without taking a prompt and absolutely unyielding Christian stand against it. He certainly considered taking such a firm Christian stand as his Christian duty. Duhem steadfastly refused to worship other men — regardless of their social importance — and he studiously avoided behaving like a hypocrite by taking care never to pay the expected visits of homage to persons in positions of importance. When he encountered what he considered not to be true, he fearlessly spoke out for the truth, or what he thought was the truth, regardless of the social importance or institutional power of the person he had to contradict. He did it even on occasions when he knew very well that as a result of his striking out against this error his professional career would come to grief. Another remarkable Christian characteristic of Duhem was the fact that all his great professional efforts were not undertaken by him for the sake of gaining prestige, but rather for the purpose of winning a wider audience and a bigger possible influence for his scientific work, which he believed was a better approximation of the physical truths than the works of others.

Duhem, who was the oldest of four children, was privately educated until he was eleven years of age. After that he attended the *Collège Stanislas*, a catholic lycée, where he proved himself to be a brilliant student. He was proficient in every subject: classical languages, modern languages and literature, history, science and mathematics. Instead of the *École Polytechnique* as suggested by his father, he decided to enter the *École Normale Supérieure* where he was placed first in the entrance examination, considerably above the rest of the candidates. He entered the *École Normale* in 1882 after some delay caused by ill health, and he passed his studies with distinction in the *licence* and *agrégation*. He published his first paper on the application of the thermodynamic potential to electrochemical cells in 1884 whilst he was still a student. In the same year Duhem presented his doctoral thesis on the subject of thermodynamic potential in chemistry and physics, even before receiving his *licence* (the French undergraduate degree) — and this is where all his troubles started, for his thesis demolished the erroneous principle of maximum work as a criterion of spontaneous chemical reactions. This principle had been the contribution of Marcelin Berthelot (1827-1907), an important functionary in the French educational establishment, who became the Minister of Public Instruction in France in 1886-1887. Duhem had been inspired by the researches on the thermodynamic potentials of F.J.D. Massieu in 1869, of Josiah Willard Gibbs in 1875, and of Hermann von Helmholtz (1821-1894) in 1882. Berthelot, who resented Duhem's demolition of his faulty principle, wielded so much influence that he was able to make the doctoral committee reject Duhem's thesis. Duhem, as a Christian, defiant of secular authority, published his thesis as a book, entitled *Le potential thermodynamique et ses applications à la mécanique chimique et à la théorie des phénomènes éléctriques*, (Paris), (1886), because he believed it to be one of his foremost Christian obligations to stand up for the truth, and to correct error and fight falsehood wherever it was to be found without regard to personal cost. After this act of defiance, Berthelot promised to see to it that Duhem would never teach in Paris, being influential enough to back up his threat until 1900. In *Le potentiel thermodynamique*, Duhem demonstrated his precocious vituosity by being able to give a logically connected exposition of thermoelectricity, pyroelectricity, mixtures of perfect gases and liquids, capillarity and surface tension, heats of solutions and dilutions, solutions in gravitational and magnetic fields, saturated vapours, dissociation, freezing points of complex salt solutions, osmotic pressures, liquefaction of gaseous states, electrochemical potential for electrified systems, stability of equilibrium, and generalisation of the Le Châtelier Principle.

This very first book by Duhem was indeed quite a remarkable piece of work of a young man of twenty three. As is the case with many extraordinarily talented minds, Duhem had already hewn out the general direction of his future research when he was still a very young man in high school. He had been inspired by his teacher Jules Moutier to study the thermodynamic researches of J. Willard Gibbs and Hermann von Helmholtz that employed two free energy functions closely related to the characteristic functions of François J.D. Massieu which Massieu had used in 1869 and 1876. Duhem called Massieu's functions "thermodynamic potentials". Gibbs and Helmholtz, who belong to that select group of men of genius — whom the nineteenth century West produced in an unusual abundance — exercised a lasting influence on Duhem's future researches. Duhem's scientific work was launched in his concerted effort to develop, broaden, and deepen the pioneering ideas of Gibbs and Helmholtz on the thermodynamic potential.

Because of the bureaucratic corruption in the French state educational institutions, Berthelot's professional persecution of Duhem was successful, and Duhem was forced to present another doctoral thesis on the mathematical theory of magnetism in 1888; this was under the jurisdiction of the mathematical faculty, and less susceptible to the conniving influence of Berthelot. In 1885 Duhem was placed first in the competitive examinations for teaching physics. Subsequently he taught physics at the university of Lille from 1887 to 1893, of Rennes from 1893 to 1894, and of Bordeaux from 1894 to 1916. Duhem was an enormously talented teacher and he was universally admired by his students, because his inspiring and clear lectures were very carefully and logically constructed, being prepared from the viewpoint of how much can be learned, rather than how much can be taught in a lecture. Duhem's lectures, as a consequence, were systematically composed to benefit the student rather than to serve the vanity and convenience of the lecturer. For these reasons, his lectures were massively attended by students. It has been said that his reputation as a naturally gifted teacher became almost legendary in his own time. Even today his autographed lectures, which are systematically interspersed with historical and philosophical remarks, are very impressive.

Right from the beginning, Duhem considered the mechanical and formalised quantitative theories of physical and chemical sciences as woefully inadequate. In his opinion the qualitative properties of physical systems also have to be considered in a sound physical theory. Duhem had always been interested in the philosophy of science, because he constructed his own scientific theories from fundamentals, but he worked especially intensively

INTRODUCTION

in this field from 1892 to 1906, and it culminated in his justifiably renowned tractate, entitled *La théorie physique, son objet et sa structure*, published in 1906 (translated into English in 1954) after he had completed most of his basic researches in physical science. To Duhem, a physical theory "is a system of mathematical propositions, deduced from a small number of principles, which has the object of representing a set of experimental laws as simply, as completely, and as exactly as possible". He was a proponent of clear and abstract physical theories that are logically integrated, consistent, and mathematically precise. He considered physical theories to be representations of physical phenomena, but not explanations of the underlying ultimate reality, the so-called metaphysical reality. He believed that abstract laws and concepts can at best merely approximate experimental observations, and he demonstrated convincingly that there is no such thing as a genuine crucial experiment (*experimentum crucis*) for testing the truth of any particular single hypothesis of a theory. Therefore no unique set of hypotheses can be determined by induction from experiments, and, hence, there may exist other sets of hypotheses which can also represent the same phenomena. It is for this reason that a great deal is left to the judgment of the scientist, which implies that theories depend upon personal taste; and taste depends upon the personal culture of the individual scientist. As the choice of the hypotheses for any scientific theory is extra-logical and presided over by the taste of the theoretician, the guide to the construction of the foundations of science, according to Duhem, is the history of science. Duhem advanced sound arguments to show that the formalised quantitative methods of science are quite inadequate for physical sciences even, and that the laws and conclusions of experimental sciences cannot reveal directly the underlying ultimate nature of things. Duhem asserted that man needs faith in his imaginative powers to guess the nature of the reality lurking behind the appearance of phenomena. The famous German poet, novelist, statesman, philosopher, and scientist, Johann Wolfgang Goethe (1749-1832) had already insisted that man needs faith to enquire about nature, because before he starts his enquiry he must believe that nature is understandable by man. In this regard Duhem wrote, "The study of the method of physics is powerless to disclose to the physicist the reason leading him to construct a physical theory". Goethe thought that the highest achievement of man in his enquiry about nature is his realisation that what is factual in nature is already theory. Goethe implied by this that the theories man has to invent, in order to enquire about the inner reality of nature, make the facts. Goethe opined that a thinking man already errs, particularly when he asks for cause and effect, for both together constitute the invisible phenomena.

Xenophanes of Kolophon (*c.* 570 - *c.* 478 B.C.), an ingenious ancient Greek thinker, had expressed views that are strikingly modern. Xenophanes maintained that man's knowledge is a web of guesses, that man can never be certain if he is in possession of the absolute truth, not even when he has it in his grasp; that what man actually has are fancies which can be taken to be something like the truth. Xenophanes emphasised that the gods did not reveal all things clearly to mortal men, but that men in the course of time can find them out through enquiry. In other words, to a great extent man constructs reality out of his imagination through enquiry, but, according to Duhem, this imagination must be disciplined by reason, critical historical perspective, and sound personal scientific experience. Duhem firmly believed that ontological questions lie beyond the province of scientific methodology, that science is always restricted to dealing with sequences of phenomena, and that it can never, with certainty, reach beyond phenomena to the underlying reality of nature, which is metaphysical. Duhem, like the Mediaeval scholar John Dons Scotus (1270-1308) and his disciple William of Ockham (1280-1348), considered faith and reason to be independent faculties of man, for reason can never prove faith—nor the other way round. Duhem thought, like Dons Scotus, that the will of man had primacy over his intellect, that the intellect was subservient to the will—and not the other way around. Following Ockham, Duhem was convinced that through sheer logic man can never directly discern the ultimate metaphysical reality behind the appearance. He followed Nicholas of Cusa (1401-1464), the last Mediaeval philosopher, who thought that human knowledge is based on man's ideas of things, rather than the things themselves, that ideas are not ready-made by nature but rather man-made, that the mind knows only its own content, its own ideas, which are conjectures and constitute mental constructs; and only through these mental representations does man understand phenomena. Duhem relied heavily on the metaphysical belief when he thought that man is capable of discriminating between rival theories and of determining which one may have a better correspondence in a certain definite respect to the perceptual manifestations of the phenomena. Furthermore, Duhem stressed another important metaphysical point concerning progress in scientific theory: no progress of physical theory is possible if man does not believe that the theory which corresponds better with the physical manifestation of the phenomena somehow reflects the ultimate physical reality of the phenomena better than the rejected theory. Duhem advanced a further metaphysical judgment, his belief that if man continues inventing rival theories about a phenomenon, and continues selecting the theory which corresponds the better with the manifestation of the phenomenon, then this process of successively improving theories leads asymptotically to a limiting form of the theory for this phenomenon that is completely unified and perfectly logical, and that marshals the experimental laws into an order that is analogous to, but not necessarily identical with, the supreme transcendent order according to which the underlying metaphysical realities are *classified*. To Duhem, a metaphysical belief in an order transcending physical science was the sole justification of physical theory. He wrote: "..., the physicist is compelled to recognise that it would be unreasonable to work for the progress of physical theory if this theory were not the increasingly better defined and more precise reflection of a metaphysics; the belief in an order transcending physics is the sole justification of physical theory". Duhem called such a limiting form of the theory for any phenomenon its "natural classification". In his famous *Notice sur les titres et travaux scientifiques de Pierre Duhem*, (Bordeaux), (1913), Duhem categorically stated: "Physics is not capable of proving its own postulates, nor does it have to prove them". Elsewhere Duhem explained: "To the extent that physical theory makes progress, it becomes more and more similar to a natural classification that is its ideal end. Physical method is powerless to prove that this assertion is warranted, but if it were not, the tendency that directs the development of physics would remain incomprehensible. Thus to find the title, to establish its legitimacy, physical theory has to demand it of metaphysics". "The more that theory is perfected, the more we apprehend that the logical order in which it arranges experimental laws is the reflection of an ontological order".

To Duhem, physical theory can only represent, but not explain, phenomena. According to Duhem, "to explain (*explicare*) is to strip reality of the appearances covering it like a veil, in order to see the bare reality itself. The observation of physical phenomena does not put us into relation with the reality hidden behind the sensible appearances, but enables us to apprehend the sensible appearances themselves in a particular and concrete form. Removing or tearing away the veil from these sensible appearances, theory proceeds into and underneath them, and seeks what is really in bodies.

"When a physical theory is taken as an explanation, its goal is not reached until every sensible appearance has been removed in order to grasp the physical reality ... it follows that in order to judge whether or not a set of propositions constitutes a physical theory, we must enquire whether the notions connecting these propositions express, in an abstract and general form, the elements which really go to make up material things, or merely represent the universal properties perceived.

"For such an enquiry to make sense or to be at all possible, we must first of all regard as certain the following affirmation:

"Under the sensible appearances which are revealed in our perceptions there is a reality distinct from these appearances.

"This point granted—and without it the search for a physical explanation could not be conceived—it is impossible to recognise having reached such an explanation until we have answered this next question: What is the nature of the elements which constitute material reality?

"Now these two questions—Does there exist a material reality distinct from sensible appearance?, and, What is the nature of this reality?—do not have their origin in experimental method, for this is acquainted only with sensible appearances and can discover nothing beyond them. The resolution of these questions transcends the methods used by physics; it is the object of metaphysics.

"Therefore, if the aim of physical theories is to explain experimental laws, theoretical physics is not an autonomous science; it is subordinate to metaphysics."

To try to establish the nature of the ultimate reality behind phenomena by inventing atomic models for matter and assigning them ontic reality, meant to Duhem subordinating physics to metaphysics and denying the autonomy of physical science. Therefore, Duhem insisted that a valid physical theory should not contain metaphysical assumptions about the ultimate inner nature of matter, but rather that it should have an abstract phenomenological form. He retained this view of physics until the day he died. For this reason Duhem opposed corpuscular theories of his day— which was not as unsound as it may seem today—because his contemporary promoters of the atomic theory went about constructing atomic structures of matter, with great arbitrariness and lack of logical consistency, in which the atoms represented hard or elastic spheres, whilst insisting that these hypotheses actually reveal the underlying ultimate reality of matter. Duhem rejected this view as illusory. A scientist such as Duhem, who had an unrivalled scope of knowledge in physical sciences, could have foreseen that the particle approach to the ultimate reality of matter leads to an infinite pursuit, because the ultimate particles will ultimately turn out to be pure creations of man rather than creations of physical nature. Soon the 'indivisible spherical atoms were subdivided into subatomic particles, thus completely scrapping the idea of the ultimate nature of atoms. Then the massive nucleus of the atom itself was subdivided into nuclear particles, and this kind of fundamental particle hunt, which has continued ever since, has produced an immense multitude of fundamental particles, and even of anti-particles. The supposed nature of these fundamental particles has changed dramatically since Duhem's day, and the theories treating their interaction begin to resemble more and more the Platonic theory of "save the phenomena".

Duhem also rejected the special relativity theory because it violated common sense, and mangled classical mechanics in order to leave the atomic theories of electrons and Maxwell's theory intact.

The development of the atomic theory and the morass of particles the nuclear theory has produced to date, lend support to Duhem's view that such metaphysical model-making and particle hunting research in physics cannot lead to the revelation of the ultimate inner nature of matter, as the naive realism of the proponents of this type of atomic physics leads them to think. Today the atoms of Duhem's time are no longer indivisible, and it is not unreasonable to assume that the number of subatomic particles will increase without limit as long as particle hunting continues, because nature is infinitely more complex that the simple ideas men can invent about its ultimate inner structure. Duhem considered any fundamental element in physical science to possess only a provisional and relative status with respect to any stage of the scientific development of physical theory. He believed that man's mind is capable of learning something about the true inner nature of the physical world, but man is not able to strip appearance from phenomena and gain direct knowledge of the ultimate nature of things. Even to extract some indirect knowledge about the deep inner nature of the physical world, in Duhem's opinion, is not possible for men if they use only quantitative methods. Duhem insisted that some qualitative considerations are also required. Physical science has veered between these two extremes, the purely qualitative methods of the Aristotelian physics and the purely quantitative methods of contemporary physics. In both extremes it was physical science that suffered, because it became stagnated as a result of any such a one-sided approach to science. It has often been pointed out that the present day 'mania' for quantification, which has penetrated virtually every compartment of human enterprise and experience, and is often presented by the dogmatic promoters of this 'mania' as the revealed ultimate truth, presents a clear threat to Western science and civilisation that cannot be underestimated. The dogmatism and the air of infallibility of the quantifiers represent a real and present danger to Western man, for it brings with it his dehumanisation and thingification, with the accompanying loss of individual freedom and free enquiry.

Duhem was quite aware of the dangers of this type of scientism, a brain child of the Enlightenment, a period which to Duhem was the age of the rebirth of paganism in Europe.

From the contemporary viewpoint, the history of science seems to indicate that no single scientific method could have been used to gain such a wide variety of results that has been achieved up to today. It appears that different methods and different philosophies have been successful at different times and for different problems. Duhem's methodology seems to be valuable for modern continuum mechanics, as in this field the researches follow essentially Duhem's general ideas of a rational continuum theory.

Duhem considered the mechanical explanation of physical nature as the most serious obstacle to the development of theoretical physics, because it was riddled with fallacies and logical contradictions. Although Duhem admired the genius of Maxwell, he was extremely critical of Maxwell's electromagnetic theory because it lacked a coherent and comprehensive logical structure, and even in its ultimate abstract form it showed signs of being a product of mechanical model-making. Duhem rejected the mechanistic viewpoint of physics, and in his own researches he consciously avoided all corpuscular theories which depended on mechanical models. Therefore, no molecules or atoms may be found in his mature scientific work. Instead of trying to reduce every branch of physical science to classical mechanics, Duhem considered classical mechanics itself as a limited case of a general rational mechanics, which he considered to be a generalised thermodynamics. He made a remarkable attempt to unite mechanics and thermodynamics into a general rational mechanics having the form of a continuum theory, and he tried to recast the theoretical foundation of physical chemistry within a generalised thermodynamic framework. However, when he attempted to do the same for electrodynamic theory he ran into great difficulties.

Duhem was a believer in abstract, comprehensive, logically consistent, analytical theories which are descriptive and practical, and are governed by a few fundamental axioms from which many physical laws can be rigorously deduced as general theorems. In this aim Duhem was a pioneer, and all work done in contemporary continuum mechanics essentially follows this basic idea of Duhem, inasmuch as his methodology is especially valuable for the theories of gross phenomena.

The generalised science of thermodynamics that Duhem tried to construct had a great analogy with the cosmology of Aristotle in which the "natural place" in Peripatetic Theory corresponds to the "entropy" in the generalised thermodynamic theory. Thus to Duhem the modifications in his generalised thermodynamics were "all the modifications of states of bodies, as well as the modifications of relationships in the variations of physical qualities", which are quite similar to Aristotle's idea of "motion" or "movement". According to Duhem this comprehensive idea of movement, which is central to his science of generalised thermodynamics, includes changes of place as well as all kinds of physical and chemical modifications, "... dilatations and contractions which alter the density; fusion, vaporisation, which modify the physical state; solutions which mix bodies; reactions which combine the elements or dissociate the compounds; phenomena of all kinds which change electricity and magnetism".

The ingenious Scottish engineer and scientist William John Macquorn Rankine (1820-1872) had already stumbled on the idea of this phenomenological science as a continuum theory in his memoir, *Outlines of the Science of Energetics*, in 1855, and he had called it 'energetics', a name that Duhem promptly adopted. The German scientist Hermann von Helmholtz had contributed to the establishment of such a phenomenological science through his efforts to find a common foundation in the continuum theory of mechanics, thermodynamics, electrodynamics, and electromagnetism. The German physicists Ladislaus Natanson and Georg Helm were also doing some research on the general thermodynamic theory at the time. The experimental laws on which the science of energetics was to be erected had been established by Sadi N.L. Carnot, Robert J. Mayer, James P. Joule, Rudolph Clausius, William Thomson (later Lord Kelvin), and Herman von Helmholtz, and had been analytically formulated by Rudolph Clausius, François J. Massieu, and von Helmholtz. Therefore, it remained for Duhem to formulate the general theorems in order to incorporate all the experimental phenomena into a science of energetics, and then apply them to the different disciplines of physics. One of the most important parts of this formulation was to establish the equations of the thermodynamics of bodies in motion, and this Duhem successfully accomplished. When he attempted to incorporate electromagnetic theory within the scope of energetics he was less successful. Duhem stated that his energetics "presents, as one can see, very clearly and very deliberately the character of a *logical system*, constructed in a formal manner and destined to *represent* phenomena. At the base of the system are abstract hypotheses, and the theory is verified only by the accord of its consequences with the facts".

In 1891-1892, Duhem published in three volumes his lectures, *Leçons sur l'éléctricité et magnétisme*, wherein he tried to give a comprehensive logical structure to the theory of electromagnetism by basing it on von Helmholtz's theory of electrodynamics. Duhem described the Maxwell theory, which he opposed, and the Helmholtz theory as follows: "... the electric current is the manifestation of a certain state of stationary movement of an unknown nature, which has its seat in the conductors traversed by the current and also in the medium which surrounds this conductor. The electrodynamic energy is the kinetic energy of this movement; to this energy one can apply the laws of dynamics (the equations of Lagrange, the Principle of Least Action, *etc.*).

"Another idea has directed the researches of M. H. von Helmholtz. At the basis of Helmholtz's theory we no longer find the hypothesis that the current is a movement, that the electrodynamic energy is kinetic energy, that the equations of dynamics ought to include the equations of electrodynamics. The intimate nature of electric current is left completely undetermined; the study of current is founded on a certain number of experimental facts and physical hypotheses, in conformity with the method followed by Ampère for establishing the law of electrodynamic force and F.E. Neumann for finding the formula of the electromotive forces of induction. One obtains thus formulae more general than that of Maxwell, in which it suffices to particularise some constants in order to obtain the formulae of Maxwell".

Duhem extended, elaborated, and improved Helmholtz's electrodynamic theory, and the resulting Helmholtz-Duhem electromagnetic theory is indeed more general than Maxwell's theory, for it contains two additional parameters; these can be assigned certain limiting values in order to produce the electromagnetic theory of light, a representation of Heinrich Hertz's experiments, and the Maxwell equations. However, the treatment of electromagnetic waves by the Helmholtz-Duhem theory ran into great difficulties, and success in that was only wrested from the theory at the expense of laborious complications and even some logical inconsistencies. As the Helmholtz-Duhem theory did not win as widespread support as he had hoped, Duhem published in 1902 a detailed historico-critical work on Maxwell's electromagnetic theory, entitled *Théories électriques de J. Clerk Maxwell: Étude historique et critique*, in which he pointed out minor mathematical mistakes, contradictions, unrigorous logical procedures and the lack of an experimental foundation. This work characterised what Duhem had written elsewhere: "To give the history of a physical principle is at the same time to make of it a logical analysis". Duhem soon realised that Maxwell's theory was gaining wide recognition amongst the physicists of his day, but he still hoped that in the future "it will be recognised that the electrodynamic work of Helmholtz was truly a splendid accomplishment and that we have done well to have held to it. Logic can be patient, for it is eternal". He expressed this hope as late as in 1913 in his *Notice*.

However, Duhem's works were not discussed, or accepted, or even read by the profession, because of the conspiracy of silence engineered by Duhem's academic enemies, such as Berthelot and his close friend Gabriel Lippmann, Henri Le Châtelier, Jean Perrin and many others. As a consequence of this professional discrimination, Duhem's contributions to science were "silenced to death", and Duhem himself was ostracised and exiled to the provinces for the balance of his professional life by these men of science, who were intellectually inferior to Duhem but occupied such influential positions in France's centralised educational bureaucracy. This stands as an outstanding example of the evils of a centralised statist educational system, the excesses of which Duhem fought with brilliant scientific and scholarly research of a scope and depth that is unrivalled since Euler.

Duhem's contributions to the general thermodynamic theory are first rate. In 1887, he wrote an important critical analysis of Gibbs' famous memoir on thermostatics, entitled *Étude sur les travaux thermodynamiques de J. Willard Gibbs*, which contains the first precise definition of reversible processes. In this analysis Duhem demonstrates that the reversible process between two thermodynamic states is made up of the limit set of real processes produced by letting the imbalance of forces between the system and the surroundings approach zero at each step. This limit of this set, with forces in balance at every step, is a set of equilibrium states. This limit is evidently unrealisable for a system undergoing thermodynamic change. This limiting process is now called a 'quasi-static' process. If the same process is carried out in the reverse order and the same unrealisable limit is reached, then the process is called reversible.

One of Duhem's capital contributions to thermodynamics was to develop its theory for bodies in a state of motion. Before Duhem's researches, thermodynamics was essentially confined to statics. From 1892 to 1894 Duhem published his famous series of memoirs entitled *Commentaires aux principes de la thermodynamique*, which he considered one of his more important contributions. In it, Duhem for the first time defines precisely an irreversible "quasi-static" thermodynamic process, a process such as hysteresis, for which the limit set of equilibrium states in one direction is not the same as in the reverse direction between the same two thermodynamic states. This paper contains two trailblazing contributions by Duhem. He gives a detailed exposition of the second law of thermodynamics, entropy, and the thermodynamic potential, and an axiomatic treatment of the first law of thermodynamics in which the toal energy (*oeuvre*) of the system and the work (*travail*) are considered to be undefined primitive quantities. The axioms of the total energy consist of the independence of the path, additivity along the path, commutativity, associativity, conservation, *etc.*. In this memoir the quantity of heat was for the first time defined abstractly in terms of work and energy. Constantin Carathéodory and Max Born made similar contributions much later, in 1909 and in 1921 respectively. With this pioneering work, Duhem became the founder of the axiomatisation of physical theory, and his influence on axiomatic research in other disciplines of science was considerable. Even David Hilbert (1862-1943), the greatest mathematician of his or our time, was directly influenced by Duhem's axiomatic work on thermodynamics to start his own great researches on the axiomatic foundations of geometry, and, later, of physics.

In 1896, Duhem published a treatise, named *Théorie thermodynamique de la viscosité, du frottement, et des faux équilibres chimiques*, (Paris), in which he investigated the actions of viscosity, which correspond to the "passive resistances" of Gibbs and belong to reversible processes, and of friction which belongs to irreversible processes. Duhem made friction a general physical phenomenon and employed it for his study of the problem of "false equilibria" within his theory of mechanical chemistry. Duhem's far reaching and significant researches in chemical thermodynamics were collected in the four volume treatise *Traité élémentaire de la mécanique chimique*, (Paris), (1897-1899), and the one volume work, *Thermodynamique et chimie*, (Paris), (1902), (second edition), (1910); English edition by G. Burgess, (New York), (1903).

Duhem's theory of "false equilibria" led him to an acrimonious polemic from 1896 to 1910 with the proponents of the "infinitely slow reaction rate" theory. The "false equilibria" viewpoint was of great help to Emile J.C. Jouguet (1871-1943), a distinguished disciple of Duhem, who was the main contributor to the theory of explosives in 1917. Even though both theories lead to identical results, to this writer the very idea of "infinitely slow rate" smacks of fragile logic. In a series of articles from 1893 to 1894, entitled *Dissolutions et mélanges*, Duhem presented a pioneering study of heterogeneous and mixed continua. From 1896 to 1902, he studied the thermodynamics of non-reversible, "quasi-static", plastic deformations and irreversible processes, involving hysteresis and creep of materials, in the memoir *Sur les déformations permanentes et l'hystérésis*; but his results, which are of a qualitative nature, were tentative in these very difficult and still very much open problems.

In 1898, Duhem gave the first explicit general proof of the Gibbs' Phase Rule, which went beyond the intensive variables by including masses, in his English language paper *On the General Problem of Chemical Statics*.

Quite significant research was done by Duhem in hydrodynamics, elasticity and acoustics. In 1891 he published his lectures entitled *Hydrodynamique, élasticité, acoustique*, (Paris), in two volumes, which exercised great influence on the profession, insofar as it emphasised Hugoniot's work on wave propagation and pioneered finite elasticity. Duhem published a further book on hydrodynamics entitled *Recherches sur l'hydrodynamique*, (Paris), in two volumes in 1903-1904, which contains some pioneering researches on the Navier-Stokes fluids, propagation of waves in viscous, compressible fluids with considerations of stability and thermodynamics. Duhem derived some rigorous theorems for "shock waves" in Navier-Stokes viscous fluids which he called "quasi-waves". A "quasi-wave" consists of very thin fluid layer with very rapid, smooth changes of properties across it. Duhem introduced viscous stresses by assumptions extraneous to thermodynamics, a practice still followed. He proved that no high order discontinuity can be propagated through such a viscous fluid, a comportment quite contrary to the behaviour of non-viscous fluids. He showed that the only discontinuities that can exist in viscous fluids are transversal. Duhem established several theorems on shock waves, some of which have been improved. Recently more general constitutive theories of fluids have been established which admit propagation of true shock waves. He generalised, completed, and emended earlier results on the stability of floating bodies, a problem first studied with incomparable brilliance by the Hellenistic engineer-mathematician Archimedes of Syracuse (287-212 B.C.).

In 1893, Duhem took a great step forward in rational continuum mechanics in an article, *Potentiel thermodynamique et pression hydrostatique*, in which he proposed the concept of an oriented continuum which not only consists of points but also possesses directions associated with each point of the continuum. Duhem's concept of the oriented body was developed by his disciples, the astronomer Eugéne Cosserat (1866-1931) and his brother, the engineer François Cosserat, in the so-called Cosserat theory of deformation of rods and shells in 1907-1909. In recent times liquid crystals, bodies with microstructure, and some dislocation problems have been treated by the theory of oriented bodies.

In his own time, Duhem was virtually a lone cultivator of a strict, rational, finite theory of elasticity in which he stressed the importance of establishing rigorous and general theorems. Duhem carried out extensive research on the rational theory of elasticity, and he published his results collectively in his treatise, *Recherches sur l'élasticité*, in 1906, which represents a repository of rational continuum theory of finite elasticity, and as such exercised considerable influence in this field of research, especially before the First World War. Duhem was the pioneer investigator of wave propagation in elastic, heat-conducting, and viscous continua undergoing finite deformations. He demonstrated that in such crystalline or vitreous elastic continua no real waves can be propagated, and that the only possible discontinuities that can occur always separate the same particles. Duhem did not investigate the possible quasi-waves of viscous solids, but he did study at length the relationships existing between waves in isothermal and adiabatic non-viscous elastic media subject to finite deformations.

Duhem was virtually the only scientist of his time to examine the general stability conditions of elastic solids, but in order to prove some general theorems he had to impose particular conditions of stress and strain. The elastic solids which he investigated did not satisfy conditions as general as are used in contemporary research on this topic.

Throughout all his productive life Duhem returned to the problem of the stability of thermodynamic systems. He attempted to generalise Gibbs' theory of thermodynamic stability by considering global stability, but this generalisation made his problems exeedingly difficult. Despite this great added difficulty he was often able to derive precise sufficient conditions for stability, but he had troubles with the necessary conditions. Whatever success he had here, he bought it by introducing special hypotheses into his stability theory.

In 1911, Duhem summarised his life-long effort to construct a corpus of physical science, the so-called energetics, as a rational phenomenological continuum theory which eschews metaphysical assumptions about the ultimate inner reality of matter, in a two volume treatise *Traité d'énergétique ou de thermodynamique générale*, (Paris), which he considered his most lasting contribution, topping, as it did, almost thirty years of intensive research in the physical and chemical sciences. This treatise also is symbolic of the most outstanding failure in Duhem's life-long efforts: the omission of the topic of electromagnetism; this implies that in his critical scientific opinion, he had failed to find a satisfactory energetical theory of electromagnetism.

Duhem is the founder of the modern history of science, and in this field he towers above all other science historians to this very day—no one in this field of scholarship has been able to come anywhere near the depth and breadth of Duhem's achievements. His fellow historians of science, in comparison with Duhem, seem amateurish, for they lack the almost universal talents of Duhem— who was at the same time a leading scientist and philosopher of science—of evaluating, analysing and criticising in a competent and profound manner the contents of the past works of science. Duhem himself stated that to criticise any work of science means to analyse and evaluate its logical structure, its hypothetical content, and its correspondence with phenomena. This can be done with complete competence and supreme confidence only by a scientist who himself has created fundamental science, who at the same time is a philosopher of science of the first rank, and a polyglot of classical and modern tongues.

Over the last two centuries, Duhem, as a historian of science, is the only man who has eminently satisfied all these requirements. It is for this reason that Duhem's historical work is so superior in comparison with the works of other historians of science.

Duhem's interest in the history of science stemmed from his creative research in science. He realised early that to effectively create new science requires the creator to understand the philosophy of science critically. In order to understand the philosophy of science and the continuity of scientific ideas correctly, Duhem had to undertake a profound and extensive study of the history of science. At first Duhem studied the history of science primarily to support his personal philosophy of science, and his investigations in the philosophy of science he carried out to support his scientific research. Therefore, Duhem was not merely a scientist, he was a scientist-philosopher-historian, a very superior intellectual. It was the intention of Duhem to try to avoid any *extreme* and *naïve* positions in his interpretation of science, and to present a balance perspective of science to his fellow men as *his* cultural contribution to his Western civilisation. Duhem was especially keen on discovering the historical precedents of the formal abstract approach to science, because this approach represented the very essence of his own methodology of energetics.

The opinion has often been voiced that Duhem discovered the history of Mediaeval science virtually single handed, which is true. But he also did profoundly significant and original research on seventeenth century physics and ancient physics. Duhem rejected the popular opinion of his day which asserted that science undergoes revolutionary changes in the course of its history, and as a result of his own firm conviction that science is always continuous, and thus evolutionary, he set out to prove that Galileo's ideas were evolved out of much earlier scientific work and were not as original as Galileo himself and others believed. To accomplish this great goal, Duhem began his historical investigations with the study of the origins of statics, the result of which he published in the two volume monograph, *Les origines de la statique*, (Paris), (1905-1906), a historical account in which Duhem traces the development of the principles of statical equilibrium from the ancient Greeks to Lagrange. In this great work Duhem had made the observation that modern science was born in the Middle Ages about A.D. 1200, and that many parts of it were plagiarised in the fifteenth and sixteenth centuries by a bevy of mathematicians who published that material as their own contributions. Duhem condemned such intellectual corruption, and emphasised that respect for tradtion is essential for genuine scientific progress.

After this revealing historical research Duhem set out to establish the actual record of the history of physical science, and it turned into the most gigantic individual achievement of all ages in the study of the history of science. Duhem began his historical study with the notebooks of Leonardo do Vince, with Leonardo's sources, and with the works of the sixteenth century scientists who learned their physics, which was actually of Mediaeval origin, from the humanists of the Renaissance. The humanists had made it a pompous habit to complain about the backwardness of the Middle Ages whilst at the same time quoting verbatim from the Mediaeval manuscripts of science which were the very source of the humanists' knowledge.

In 1905-1906 Duhem published the results of this epoch-making and profound research in the three volume historical monograph *Études sur Léonard de Vinci: Ceux qu'il a lus et ceux qui l'ont lu*, (Paris), which convincingly revealed to Duhem the abundance of the sources of Mediaeval science. He then undertook what is certainly the most monumental systematic study of the history of science by one man—the history of physical theory from the Ionian natural philosophers to the rise of classical physics. The astounding amount of path-breaking historical research which Duhem did in such a short time is unequalled to this day. He was able to prepare ten manuscripts for a planned twelve volume work in four years, and see the first four volumes off the press between 1913 and 1916, and put the fifth volume, which appeared in print posthumously in 1917, through production. Duhem's daughter, Hélène Pierre-Duhem, supervised the publication of the last five volumes of the *Système du monde* between 1954 and 1959.

As Duhem was a man of high culture, firm convictions and definite tastes, all his judgments were in conformity with his basic views. He was an unabashed and enthusiastic French nationalist, and his patriotic feelings influenced his value judgments, with the effect that he tended to overestimate the achievements of French scientists and to underestimate the contributions of scientists from other nations. He especially underestimated the contribution of British scientists, and he classified their minds as broad but shallow because of their laxity towards logical rigour and their lack of interest in systematic mathematical theories of science. In particular, he did not give enough credit to the scholars of Merton College in Oxford for their contributions to the theory of the kinematics of motion, nor to Thomas Bradwardine (*c.* 1290-1349) for the significance of his reformulation of the Aristotelian law of motion. At the same time, Duhem overestimated the importance of the University of Paris in the new physics, the contributions of Jordan de Nemore (*fl.* before 1260), Jean Buridan (*c.* 1300 - after 1357), and Nicole Oresme (*c.* 1330-1382); Duhem also overestimated the certainty of the scientific philosophy of the Christian philosophers of the Middle Ages, which in his opinion reduced to the principle of the greatest simplicity and to the principle of the greatest precision. He credited the scholars of the University of Paris with the notion that the superlunary and sublunary worlds both obey the same physical laws, whereas it was actually first enunciated by the German Mediaeval scientist, Nicholas of Cusa. Even though in his monumental historical *magnum opus*, *Système du monde*, Duhem had moderated his earlier assessment of French contributions, it is undeniable that his French patriotism and his religious sympathies influenced his judgments in every one of his historical works. Of course, Duhem was fully aware that to decide on the merits of any creative work of science required a definite cultural, *i.e.*, moral, philosophical, religious, and intellectual, point of view.

In 1908 Duhem wrote a series of five articles in the journal *Annales de philosophie chrétienne* that immediately afterwards was issued in book form under the title, *Sozein ta phainomena: Essai sur la notion de théorie physique de Platon à Galilée*, (Paris), which is essentially a work of documentation of the most important historical works on physics, but especially on astronomy, the only exact science until relatively recent times. The main import of this scholarly work is that formal mathematical science has always played the most substantial rôle in the development of Western science. This small book, which was published in English in 1969 under the title *To Save the Phenomena. An Essay on the Idea of Physical Theory from Plato to Galileo*, (Chicago), to which is appended a valuable historical introduction, can be viewed as a sort of condensed panoramic view of the historical treatise *Système du monde*, regretably not translated into English.

Duhem was engaged in historical research from 1895, but he did his intense research on the history of science in the period between 1892 and 1916. His main historical treatises are:

Le mixte et la combinaison chimique. Essai sur l'évolution d'une idée, (Paris), (1902);
L'évolution de la mécanique, (Paris), (1902);
Les origines de la statique, two volumes, (Paris), (1905-1906);
Études sur Leonard de Vinci, ceux qu'il a lus et ceux qui l'ont lu, three volumes, (Paris), (1906-1913);
Sozein ta phainomena: Essai sur la notion de théorie physique de Platon à Galilée, (Paris), (1908);
Le système du monde. Histoire des doctrines cosmologiques de Platon à Copernic, ten volumes, (Paris), (1913-1959).

Duhem was the only creative scientist at the turn of the century to advance methodological doctrines in support of his physical theory, and he easily saw through the fallacy of the atomic theory of his day. Duhem was a genuine teacher whose work represented a serious effort to elucidate the ideas underlying science, and science to him was a search for the truth. It was never Duhem's motivation, in writing his great treatises on the history and philosophy of science, to heap professional prestige and academic honours on himself. It lies beyond the powers of belief of faithless modern man to accept the fact that Duhem was motivated and animated to produce his scholarly work primarily in order to live up to his deeply felt obligation to the mission of Mediaeval Christianity of helping his fellow men by "bringing more light" to his Christian civilisation. It was the profound Christian convictions of Duhem that his contemporary free-thinking French academic colleagues, most of whom had been effectively paganised through the influences of the philosophs, despised and

INTRODUCTION

could not tolerate in him.

While Duhem was openly and fearlessly declaring his views, opinions, and criticisms in public, just as Christ had taught, his unprincipled adversaries were conniving and conspiring behind closed doors in an effort to arrest his professional progress, which he so richly deserved, and to keep him in low paid teaching posts. This professional discrimination against Duhem, the most outstanding French intellect of his time, was very successful. Duhem never was allowed to teach in Paris, where he properly belonged on the basis of his overall intellectual brilliance. If anybody should have been teaching in Paris, it should have been Duhem. Duhem faced his persecution by men who were his intellectual inferiors with an admirable Christian serenity and unshakeable Christian integrity, for he never yielded to the will of these unprincipled men, nor complained about the despicable professional treatment he received from them. What is even more remarkable about him is the fact that Duhem made every effort to live up to the principle of personal Christian charity, which prevented him from becoming mean and vengeful as a result of his persecution. A clear example of his charity is evident in his treatment of Berthelot, the man directly responsible for initiating his decades-long unwarranted professional ostracism. Berthelot, as an old man, finally repented of his actions in the ostracism of Duhem, actions he assiduously promoted for nearly twenty years, and had a change of heart; this he demonstrated by ceasing his connivance against the professional promotion of Duhem, and by trying to make amends in 1900 for his former acts of ill will through voting for the admission of Duhem as a corresponding member of the French Academy of Sciences. Duhem recognised in Berthelot's positive actions a genuine repentance for the latter's former unjust treatment of him, and he immediately reciprocated with a generous and honourable personal attitude towards Berthelot, just as if the latter's persecution of him had never taken place.

Duhem was a man whose nature was proud but not pompous, sensitive but not vain, caustic in any confrontation but not vindictive, scrupulously upright but not hypocritical, self-confident but not arrogant, self-assertive but not autocratic—in other words a man worthy of genuine professional respect. Instead of respect, Duhem received from the influential section of the intellectual bureaucracy of France unmitigated envy, gross professional discrimination and ostracism—unprecedented even in France—and all because of Duhem's deep Christian beliefs, adamant political convictions, and, above all else, because of his abundant, far-reaching, and awesome scientific and scholarly talent and intellectual independence. This type of shabby professional treatment has not been confined just to France, for throughout history—but especially in modern times—the kind of treatment Duhem received has always been virtually the standard practice of intellectual life in statist countries with a centralised intellectual bureaucracy accepted by the intellectual community in the same way primitive man accepted the weather—with complete resignation. In the last century there was a similar case in Germany, where Hermann Günther Grassmann (1809-1877), an outstanding mathematical genius, was to remain a highschool teacher all his professional life, because another mathematician, Johann August Grunert (1797-1872), minute in talent in comparison with that of Grassmann, was envious of Grassmann, and who, because of his position in the educational establishment, was able to block Grassmann's appointment to the chair of mathematics at a university. In recent years, an outstanding case of academic ostracism took place in the United States, a country that has no strong centralised educational bureaucracy. The ostracism of Ludwig von Mises (1881-1973), one of the greatest economists, historians, and philosophers of any age, whose *magnum opus*, *Human Action*, ranks with Adam Smith's *The Wealth of the Nation*, was a result of the virtually monolithic economic ideology of the leading economists in the departments of all the major universities. These dominant unversity economists were promoters of the centrally controlled *command* economy, whereas von Mises was the dean of free-market economics, which is based on the philosophy of individual freedom. Even though von Mises' economic theories, which demolished the scientific basis of Marxist and socialist economics, were the only theories capable of explaining the phenomena of inflation and recession, he was totally ignored, and his works were met by a wall of silence. Ludwig von Mises fared even worse than Duhem, he never was offered a paid university teaching position. Von Mises, very much like Duhem, was a moral intransigent: he never compromised in the least on any of his moral principles. If he had done so, any choice position could have been his. But, like Duhem, he had an iron clad integrity, which made von Mises one of the outstanding characters of his age, as Duhem had been in his. Both men had to pay a high professional price for their virtue—which they were quite willing to do.

Near the end of his life, Duhem's fortunes turned for the better, for he was finally elected a corresponding member of the French Academy of Sciences in Paris in 1900 with Berthelot's help. In 1908, Duhem refused the nomination for the Legion of Honour on the basis of his Christian beliefs, because to accept this Republican accolade signed by a man whom he despised would have been hypocritical on his part. He rejected the offer to be a candidate for the prestigious chair of the History of Science at the *Collège de France* in Paris, because he refused to enter Paris through the back door of history. He said that either he would go to Paris as a physicist or not at all. In 1913, when a new section was opened for six non-resident members in the Academy of Sciences in Paris, Duhem was elected almost unanimously as one of the first such members. In this matter, Duhem again demonstrated by his actions his deep abiding belief in his Christian principles, for he tried to withdraw his candidacy so that the next closest candidate—a ninety year old natural scientist whose work Duhem admired—could be elected to the membership of the Academy of Sciences in Paris before dying. He only agreed to let his candidacy stand when he was convinced by others that his candidacy had no effect on the election of this very old naturalist.

In the support of his candidacy for the non-resident membership of the Academy of Sciences in Paris, as is required, Duhem wrote a personal assessment of his work, a valuable document on all accounts, *Notice sur les titres et travaux scientifiques de Pierre Duhem*, (Bordeaux), (1913), which was published posthumously in a special issue of *Mémoires de la Société des Sciences de Bordeaux* in 1917. In this *Notice*, Duhem gave a detailed summary of his aims, motivations, and achievements in theoretical physics and in the philosophy and history of science, and the bibliography of his works in all these areas when he was a candidate for the Academy of Sciences in Paris in May, 1913. It is a valuable source for Duhem's authentic opinions of his own work.

As stated earlier, Duhem was the only scientist at the turn of the century who advanced methodology in support of his physical theory. The majority of the physicists at that time were of the mechanical school, whereas Duhem belonged to the non-mechanical school. The mechanical theory of physics had been established by Christiaan Huyghens (1629-1695), Isaac Newton (1642-1727), Pierre-Simon de Laplace (1749-1827) and Siméon-Denis Poisson (1781-1840), and developed by James Clerk Maxwell, Josiah Willard Gibbs, and Ludwig Boltzmann (1844-1906). In the mechanical theory of physics all physical phenomena are reduced to mechanics. Duhem opposed categorically this reductionism in science, and he went even further and asserted that science itself is not the sole organ of knowledge. Duhem believed that a correct idea of science cannot be obtained if the knowledge of science is limited only to the current state, or as he stated so succinctly, if the scientist is aware only of "the gossip of the moment".

Duhem was a man of a rather short stature and he looked robust, yet actually he had never known a single day of robust health. Despite his sickly constitution, Duhem was of a frank, chivalrous, and friendly disposition, but this disposition promptly evaporated when he encountered anyone whose teaching or writing seemed in his opinion to retard the advancement of science. Then he became a caustic and devastating critic, and he considered anyone like that almost as his personal enemy. In every discipline of science in which he was active he made enemies in this way. These people suppressed his work, if they could; if they could not, then they appropriated it without credit.

Duhem's personal manners were as correct and precise as his literary style. Duhem wrote with an incredible ease, without notes or corrections, countless pages of manuscripts ready for the press. This extraordinary literary ability was one professional quality of Duhem that was essential for his enormous output of scientific literature. Duhem published twenty two books in forty five volumes, and nearly four hundred memoirs and book reviews in more than twenty different journals. Two memorial volumes on Duhem's scientific work were published by the *Société des sciences physiques et naturelles de Bordeaux* in its *Mémoires*, 7e série, tome I, entirely devoted to Duhem. It contains the following studies:

1er cahier (1917): 'Pierre Duhem' by Edouard Jordan, pp. 9-39; 'Notice sur les titres et travaux scientifiques de P. Duhem', (edited by himself, and preceded by a list of his publications up to 1913), pp. 41-169.
2e cahier (1927): 'La physique de P. Duhem', by O. Manville, pp. 173-605; 'Bibliographie des travaux de P. Duhem', pp. 607-630 (this lacks about twenty five articles and more than fifty book reviews); 'L'Oeuvre de P. Duhem dans son aspect mathématique', by J. Hadamard, pp. 637-665; 'L'Histoire des sciences dans l'oeuvres de P. Duhem', by A. Darbon, pp. 669-718.

Other bibliographical and biographical sources are:

E. Jouguet: *L'Oeuvre scientifique de Pierre Duhem*, Revue générale des sciences pures et appliquées, vol. 28, No. 2, (30 Janvier, 1917), pp. 40-49.
E. Picard: *La vie et l'oeuvre de Pierre Duhem*, (Gauthier-Villars, Paris), (1921), 44pp..
H. Bosmans: *Pierre Duhem*, Revue des questions scientifiques, (Louvain), (1921), 58 pp..
F. Mentre: *Pierre Duhem, le théoricien (1861-1916)*, Mémoires de l'Académie des Sciences, vol. 57, 2 série, (1922), pp. 49-67.
P. Humbert: *Pierre Duhem*, (Bloud et Gay, Paris), (1932), 150 pp..
H. Pierre-Duhem: *Un savant français: Pierre Duhem*, (Preface by Maurice d'Ocagne, (Plon, Paris), (1936), 256 pp..
M. d'Ocagne et al.: *L'Oeuvre historique de Pierre Duhem*, Archeion, (Roma), vol. 19, (1937), pp. 121-151.
A. Lowinger: *The Methodology of Pierre Duhem*, (Columbia University Press, New York), (1941), 184 pp..

A very useful concise account of Duhem's work, in English, is:

D.G. Miller: *Duhem, Pierre-Maurice-Marie*, in: C.C. Gillespie: *Dictionary of Scientific Biography*, vol. IV, (Charles Scribner, New York), (1971), pp. 225-233.

The present volume, *The Evolution of Mechanics*, is an English translation of *L'Évolution de la mécanique*, (A. Hermann, Paris), (1905). As was mentioned earlier, Duhem was an effortless writer of scientific French prose, and he had a style of writing that cannot be translated into English with all its particular flavour and style, but the translator has made an effort to stay as close to Duhem's style in language as is possible without making the English text incomprehensible. Translations of stylish literary works can never be accurate in all respects to the original. As Duhem himself was wont to say, *traduttore, traditore* — to translate is to betray! It is the hope of the editor that the betrayal of the author committed by this translation is minimal.

Duhem wrote this book to demonstrate that his more abstract general thermodynamic theory—the so-called energetics—is just as valid a theory of physical science as the mechanistic theory of science. The first part of this book contains a historico-critical examination of the mechanical theory of physical sciences. The second part of this book represents a critical introduction to Duhem's own theory of physical and chemical sciences—the energetics—the basic nature of which has been discussed in this Introduction.

The title of this book meant to Duhem the development of the mechanical explanation of nature from its narrow scope into a general rational mechanics.

ACKNOWLEDGMENTS

The author would like to express his indebtedness to Dr. Donald G. Miller, who many years ago provided the author with a number of photographs of Pierre Duhem which originally came from the collection of Mlle Hélène Pierre-Duhem, the daughter of Pierre Duhem.

The author is grateful, also, for the competent help of the translator, Michael Cole, who critically read the manuscript of this Introduction and suggested a number of changes in its writing which have substantially contributed to its improvement.

GUNHARD Æ. ORAVAS
McMaster University

Pierre Duhem at forty years of age as professor of theoretical physics at the University of Bordeaux.

Pierre Duhem at twelve years of age in his first year at Collège Stanislas.

Pierre Duhem as an upper classman at Collège Stanislas.

TRANSLATOR'S NOTE

Declamatory, magisterial, grandiloquent; precise, perceptive, profound—these are the qualities of the writing of Pierre-Marie-Maurice Duhem; his a style common to his era, an era in which theories were discovered rather than proposed. To translate his work with effect is to reproduce the ethos of another time; it is three quarters of a century since the second edition of *L'Évolution de la mécanique*, near eighty years since Duhem assembled his thoughts for the writing of those articles which became this book. Much has happened, and much has slipped away since then.

The new student, or the casual reader, of the history of science will find abundant interest in this book; the seasoned researcher will relish it. But it is for the former that this Note is designed, for if he is not already familiar with literature of a bygone age this work will require him to set his mind in another time and to think much more carefully about the exact meanings that can be extracted from the text. Carefully chosen words have been used for certain notions, and, where necessary, an explicatory footnote has been inserted. Duhem always used the term 'explication' instead of 'explanation', as the new reader will find is convention in serious works on the philosophy of science; to explicate means to give a carefully and cogently reasoned set of deductions derived from a number of very precisely defined axioms; to explain has become a much weaker notion, encompassing good excuses and merely some sort of reasoning, besides clear, precise deduction; it is therefore much more satisfactory for the new reader to adopt this term, along with all its deeper connotations.

The combination of the particular time at which Duhem wrote this book—recalling the theoretical developments in physics at that season—with the actual content of it—and all that it implies was in Duhem's mind—makes this work a most tantalising one to reflect upon. The development of physics has followed a course that is very different from the one imagined or expected by him at the turn of the century, a fact impressed upon the reader when he recognises names which Duhem cites for certain fields of study that are quite remote from the work for which they are remembered nowadays, or from effects that nowadays bear their name. It is the latter which causes one to realise that whilst this book was being written and published in a number of forms, during the first five years of this twentieth century, there were being written and published during those same years the quantum explanations by Planck and Einstein of the then extant respective problems of black body radiation and photo-electric emission, followed by Einstein's introduction of the Special Relativity Theory. The fashionable era of levers, springs, whirlings, honey, and plum puddings was fast approaching its close.

Every student of a physical science should read Part I of this book; it is an authoritative and masterly account of the development of thought in Natural Philosophy. It shows how various ideas became favoured, were developed, and then set aside; how others were preferred, altered, and set aside in their turn; and how returns were made to those previously rejected—and so on. It leads the reader to the general corpus of knowledge built around classical mechanics and physics that a graduate would expect to have encompassed in one way or another. Every student of a physical science should expect to have an awareness of the history presented in this Part I; he will find it deeply rewarding, certainly in later years.

The student reading Part II will find himself envelopped in a rolling survey of the developments in mechanico-physical theories at the end of the nineteenth century. It will be a world of the greatest interest, because the reader will understand more clearly how radical was the effect upon theoretical physics and mechanics of the work of Planck and Einstein published in those years during which Pierre Duhem imagined, wrote, and polished this work; for the reader will receive an impression that Duhem's great railway engine ran into a vast goods yard, from whence the various wagons of his train were redistributed to many different places, whilst somewhere further up the line a small train had edged out of a siding onto the main line, and grew gradually in size and power as it accelerated down the track of history, taking on board all those passengers, waiting and arriving at the stations ahead, whom Duhem's mighty express had expected to carry to their destinations. One wonders whether history is going to give a repeat performance for the new century approaching.

The bombast, dignity, and repose of Duhem's prose can be conveyed, today, only in a style that seems a little stilted, even archaic at times; but his vocabulary and grammar are very precise, be sure of that—what he meant, he wrote: where there is latitude of interpretation, he intended it; where there is sharpness of statement, no latitude of interpretation is allowed. Everyday English has become, in these eighty years, a more limpid language; its use in public communication is vastly more widespread than then, and over the years words have been hammered into many incorrect moulds, losing, in that process, their sharpness of meaning, and often gaining many extra unhelpful tacit connotations. It therefore seems much more appropriate to translate this book of Pierre-Marie-Maurice Duhem so that it sounds as if he were an academic of note, of his day, writing in the English of his contemporaries. In this way I believe the ethos of his time is most likely to be recreated, and in that I hope I shall betray him the least.

MICHAEL COLE

AUTHOR'S INTRODUCTION

In the middle of the nineteenth century Rational Mechanics seemed to be ensconced upon foundations as unshakeable as those upon which Euclid had assembled Geometry. Assured of its principles, possibly it was allowing the harmonious development of its consequences to slip.
The rapid, incessant, tumultuous growth of the physical sciences had come to disturb this peace and ruffle this assurance; pestered with new problems, Mechanics took to doubting the firmness of the bases upon which it had reposed, and it took again to its march towards a new development.
What route would it take? Several paths lay in sight; the entrance to each of them was wide open and quite smooth; but hardly had one gone along a path than one saw the causeway shrink, the track of the route become unclear; soon one would see no more than a narrow path half hidden by thorns, cut accross by bogs, bounded by abysses. Amongst these paths, who would want to become lost in such parched isolation, who would ever be stopped short at the edge of a precipice? Where is he who would be carried through to the end desired, who, one day, would come upon the royal way? Mechanics hesitated, anxious, she cocked her ear towards those who would pretend to guide her, she pondered their contrary advices and knew not in whom she should confide.
The director of the *Revue générale des Sciences* has expressed the desire that the state of uncertainty in which Rational Mechanics is wavering should receive an exposition for the readers of this *Revue* in a series of articles of an unusual fullness[1]; I have been done the great honour of having had this exposition committed

---

[1] These articles were published in the *Revue générale des Sciences* on the 30th January, 15th February, 28th February, 15th March, 30th March, 15th April and the 30th April, 1903. Here may it be permitted to thank M. L. Olivier for his grand hospitality.

AUTHOR'S INTRODUCTION

to me, which is the origin of this book. It is certain that this state of doubt is, for every man of thought, a worthy object of consideration; for, upon the fate of Mechanics, on the method by which she will develop her theories, depends the very form of all of Natural Philosophy.

In enumerating the various paths which, turn by turn, entreat the preference of Mechanics, in reckoning the chances which each has of leading to the solution of the problems posed by Physics, I will not pride myself in impartiality. Amongst these routes there is one upon which I have been working for twenty years, devoting all my efforts to extending it, smoothing it out, cleaning it, to render it more direct and more sure. Could I believe that they who have laid out the first track have laboured in vain, and that I could add only a useless affliction? Could I believe that Mechanics will proceed in another direction?

Impartiality, moreover, is required of a judge; but between the various inclinations which canvass Mechanics it is here not a question to decide. It is by the fruit that one judges a tree; now the tree of Science grows with an extreme slowness; centuries slipped by before it was possible to pluck ripe fruit; hardly today can we express and appreciate the quintessence of doctrines which flourished in the seventeenth century.

He who sows therefore cannot judge the value of the grain; but he must have faith in the fertility of the seed, in order that, without fainting, he may follow the furrow he has chosen, throwing ideas to the four winds of heaven.

PART ONE

MECHANICAL EXPLICATIONS

CHAPTER I

PERIPATETIC MECHANICS

At the beginning of his *Traité de la Lumière*, Huyghens defined the "true Philosophy" as that "in which one conceives of all natural effects in terms of Mechanical arguments". "What has to be done, in my opinion", he added, "may renounce any hope of ever understanding anything in Physics".

Most physicists would consent, I think, to defining the object of their science as Huyghens did; they would agree less easily amongst themselves if they had to declare what they understood by "Mechanical arguments".

Similarly, chemists in every age and in every School have thought that analysis has as its goal to determine how a body is composed from its elements; however, this would have not at all the same meaning for a disciple of Aristotle or for a pupil of Lavoisier; for a scholastic who believed all bodies to be formed from earth, air, fire and water; for an alchemist who sought in them salt, sulphur, mercury and black earth[1]; for a modern chemist who here reveals and there decides the amount of some ones of our eighty four simple bodies.

Thus, through the course of the centuries and according to the vicissitudes of the schools and systems of thought, the meaning of the words "Mechanical explication of physical phenomena" has varied incessantly, tossed about between two interpretations, opposite in the extreme, which are, as it were, the limits. The one of these interpretations comes from the powerful analysis of Aristotle; the other, prepared over a great while by the atomists of Antiquity and the Renaissance, took its finished form in the thought of Descartes. Let us sketch first of all a quick picture of Peripatetic Mechanics and Cartesian Mechanics.

By analysis of what we say, Aristotle wished to penetrate what we think, for language expresses the thought; the analysis of

---

[1] *Fr. terre damné;* this may also be translated as 'earthen residue'. (Translator's note).

thought, in its turn, is the very analysis of reality, for our reason knows that which is; the distinction of *categories* is thus at the very basis of the peripatetic system.

Against the first category, which is that of *substance* (οὐσία) are opposed the multiple categories of particularities. Amongst the particularities there are those which are not inherent to the subject in which they are encountered; such is *place* (τόπος), which depends upon the relation of a body with the bodies which surround it; but there are, against this, those which belong properly to the subject, and those latter are classed in two categories: *quantity* (πόσον) and *quality* (ποῖον).

Quantity is nicely defined by the following characteristic: every quantity of a given kind can be obtained by the juxtaposition, by the simultaneous consideration, of quantities of the same kind and of less size, and that without the order in which one considers the component quantities influence the resultant quantity; this character Aristotle expresses by saying that *quantity is that which has some parts outside others*, and the Modernists by saying that *quantity is that which is susceptible of addition*. Thanks to this character the comparison of different quantities of the same kind can always be reduced, by a sort of transposition, to the comparison of different quantities of another kind, and, particularly, to the comparison of different numbers. Consequently from this transposition, which constitutes *measurement*, the science of numbers, Arithmetic, there comes the general theory of quantity.

"*Quality*", said Aristotle, "is one of those words which are taken in many senses". Quality, the form of a geometric figure, which may be a circle or a triangle; qualities, the sensible properties of bodies, hot and cold, clear and opaque, black and white; qualities, also, but *hidden qualities*, the properties which are not apprehended directly by the senses, but from which follow certain perceptible effects: heaviness or lightness which draws a body towards the centre of the Earth or draws it away, the property of magnetism by which iron runs to a magnet.

There are some qualities that are not susceptible of being either 'the more so' or 'the less so'; a circle is not the more nor the less circular, a triangle is neither the more nor the less triangular. But the majority of qualities are susceptible of being 'the more so' or 'the less so'; like the lightness of the bowstring which the archer pulls or relaxes, they are capable of *extension* or of *remission*; a hot body can be the more or the less hot.

Between the magnitude of a quantity and the intensity of a quality there exists an essential, profound distinction. Every quantity of specified magnitude can be obtained by adding one or other various quantities of the same kind and of smaller magnitude, which are its parts. There is nothing similar in the category of quality; qualities less intense are not parts, fragments, of a more intense quality; juxtapose as many bodies as you will of which the heat intensity is that of boiling water—you will not make a body whose heat intensity is that of red-hot iron; heap up snowballs as you may, said Diderot, you will not be able to heat an oven; each degree of intensity of a quality constitutes, so to say, a breed apart; the degree of heat of boiling water is irreducible to every other degree of heat; it is not contained, like a part in the whole, in a more intense degree of heat; it cannot be fragmented into less intense degrees of heat; *the notion of addition does not enter into the category of quality.*

Amongst the particularities of which a substance is capable, there is that which actually exists for it at the moment it is considered; they are there *in actuality*; indeed, it is by this word *actuality*[1], *actus*, that the Scholastics translated the word ἐντελέχεια used by Aristotle. Other particularities, on the other hand, are not realised in the substance; they are simply possible in it; they are there *in potential*[2] say the Scholastics, who translate by the word *potentia* the word δύναμις used by the Stagyrite.

The actual state, the potential state, they do not exhaust all the states under which one can conceive a particularity; it is a third state where the potential and the intelligence are linked in a way that is inextricable as well as inexpressible; it is a state of *motion*, κίνησις.

For example, what is the melting of ice? In ice the state of water is potential; if we consider this state as purely potential, we shall have the notion of ice that might melt, not of ice that is melting. Shall we simply regard this state of water as actuality? It is then water of which we conceive, no longer ice. To conceive of the melting of ice it is necessary for us to regard the state of water as being essentially potential in ice and at the same time as laying hold of actuality.

Thus in the analysis of every motion we discover a thing that is conceived as actuality at the same moment that it is conceived of as being essentially potential. The word *motion* has as its object the expression of this intimate union between potentiality and actuality, a union to which human language cannot attempt to give expression without entering a vicious circle; for, always and perforce metaphorically, it would borrow from motion even the word by which it would try to define motion. Such is the meaning of Aristotle's celebrated proposition[3]: Η τοῦ δυνάμει ὄντος ἐντελέχεια, ἦ τοιοῦτον, κινησύς ἐστιν, which the Scholastics translated in these terms: *Motus est actus entis in potentia, quatenus in potentia est.*

---

[1] *Fr. acte;* this word is more suitably translated as 'intelligence', although in French it can mean both 'act' and 'instrument'. *Lat. actus* means 'motion' or 'putting into motion', whilst clearly here it implies the notion of a means of enabling some study to be begun. In English literature 'intelligence' is a word formerly used to connote information; and that is what the author really has in mind.
[2] *Fr. enpuissance.*
[3] Aristotle: Φυσικῆς ἀκροάσεως, Γ, α.

The example we have chosen for explaining this definition of motion, namely the melting of ice, is quite removed from what we understand by the word motion; in present usage this word denotes only the alteration of position in space; the meaning of the word is infinitely more extended in peripatetic language; without doubt, alteration of position (κατὰ τόπον μεταβολή) characterises a genus of motions, *locomotion*[1]; but even if one is restricted to considering corporeal things, one discovers a host of other motions. When a body melts, the quality that the word *fluid* (ὑγρὸν) expresses passes from potentiality to actuality; the quality expressed by the word *solid* (ξηρὸν) loses its actual state only to subsist in potentiality; and that also is a motion, but a motion quite distinct from locomotion; such a motion is named ἀλλοίωσις by Aristotle, and *alteratio* by the Scholastics.

The variety of the motions of alteration is infinite; a body which is heated or refrozen, a flame which becomes the more or the less brilliant, a piece of iron which is magnetized or demagnetized, all suffer alterations.

Locomotions, the motions of alteration, still do not exhaust the multitude of changes which are produced in the world of bodies; by these motions particularities alone are modified; now, it is changes that impinge upon the substance itself, such as those which combine elements to form a mixture, which dissociate a mixture to recover the elements. In fact, when a mixing occurs, the substances of elements lose their actual existence; in the mixture they are no more than potential; they can be recovered afresh by chemical analysis, which causes these substances to pass from potentiality to actuality; there is then *corruption* (φθορὰ) of the mixture and *generation* (γένεσις) of the elements.

Such, sketched with bold strokes, are the notions to which all physicists will reduce all the effects that bodies present; when this reduction has been done, the *explication* will have been achieved.

If one asks, for example, why lodestone attracts iron, one will reply that in the presence of lodestone the substance of the iron is *altered*, that it acquires a certain hidden quality, the magnetic property, and that the nature of this property is that of moving the iron towards the lodestone. The observations of physicists will be able to detail this explication; they will be able to specify the particular marks of the magnetic property and of the motion that it determines; but they will be able to discover nothing beyond this quality, of which it is the explication; they will be able to reduce it to nothing more elementary nor more simple, for it is the very and ultimate cause of the phenomena observed.

---

[1] *Fr. lit.* 'local motions'.

CHAPTER II

CARTESIAN MECHANICS

The renaissance of the sciences at the beginning of the seventeenth century brought a reaction against such explications; the hidden qualities were then overwhelmed with lampooning; thanks to the immortal vigour of Molière, the guffaw of laughter that they raised has carried on even up to now. It would be an interesting task, and one full of philosophical instruction, to follow the sudden turn of fortune in this context between the old Scholasticism and the new Physics; perhaps some day we shall try to bring it to a happy end; in every case it would exceed the bounds of this work.

Being directed by men who, almost all, were great geometers, the trends of the scientific renaissance found their full blossoming, and, so to say, their extreme success, in Cartesian Physics.

With Descartes, the notion of quality was banished from the whole domain of Science, which became the kingdom of pure quantity, the *universal Mathematics*.

Amongst all the sciences, Arithmetic alone was excepted from any notion borrowed from the category of quality; it alone conformed to the ideal Descartes proposed for the complete science of Nature.

In Geometry, the mind runs into the qualitative element, for this science it remains that "if kept to the consideration of figures it cannot exercise the understanding without tiring the imagination. The scrupulousness with which the Ancients used the terms of Arithmetic in Geometry, which could only proceed from this: that they did not see clearly enough their bearing, caused a great deal of obscurity and hindrance in the way in which they were explained". This obscurity, this impediment, will disappear if one dismisses the qualitative notion of figure from Geometry, in order only to preserve the quantitative notion of distance, rather than the equations that link one set of distances with another set of distances between the points that one studies. Although their objects are different, the various branches of Mathematics considers those objects which are none "other than the various ratios or proportions that exist", so that it suffices to treat these proportions in general, by the paths of algebra, without worrying about objects when they are met, or about figures when they are realised; by that, "everything that falls under the consideration of geometers is reduced to a single genus of problems which is to look for the value of the roots of some equation"; all of Mathematics is reduced to the science of numbers; one treats quantities only; qualities have no place.

With qualities expelled from Geometry, it is now necessary to expel them from Physics; to succeed in that it suffices to reduce Physics to Mathematics which has become the science of pure quantity; this is the work that Descartes is going to attempt: "I accept no principles in Physics, but that they are received in Mathematics also".

First of all, what is matter? "Its nature does not consist in hardness, neither in its heaviness, heat or other qualities of this kind", but only in "its extension in length, breadth and depth"; this is none other than matter "divisible, mobile and endowed with a figure that geometers call quantity, and which they take as the object of their proofs". Matter is therefore quantity; the quantity of a certain amount of matter, it is the volume which it occupies; a vessel contains just as much matter whether it be full of mercury or full of air. "Those who pretend to distinguish material substance from extension or quantity either put no idea into the name of substance or have confused the idea with that of an immaterial substance".

What is motion—by which I mean locomotion? It, again, is a quantity. Multiply the quantity of matter that fills each body of a system by the speed that animates this body; add together the total of these products and you will have the quantity of motion of the system; as long as the system collides with no other body which gives or takes motion from it, it will keep an unchanging quantity of motion.

Thus, throughout the whole Universe there is scattered a unique, homogeneous matter, of which we know nothing except that it has extension; this matter is divisible into parts with various figures, and these parts are mobile with respect to each other; such are the only genuine properties of that which forms bodies; all the apparent qualities which affect our senses must reduce to these properties.

Most certainly, the conception of such a Physics is admirable for its simplicity; but, by draining it of all content which would not be purely geometric, Descartes reduced it to an empty phantom, incapable of representing the world of bodies.

Cartesian matter is only "extension in length, breadth and depth". How can one conceive of such mass as being capable of motion[1], here speaking of local motion, the only one there is

---

[1] On this point the reader would consult the following work with great profit: Arthur Hannequin: *Essai critique sur l'hypothèse des*

to be in the new Philosophy?

In order that a body may be said to be in motion[1], it is necessary for it to occupy a certain place at a certain instant of the duration, and another place at another instant; one can therefore conceive of motion without conceiving of both that the place of the body has changed and that the body has remained the same. Now, what meaning can these words have if the body is identical with that part of the extension that is occupies? Could one, without being nonsensical, say that one particular part of the extension occupied successively different places? Doesn't it suffice, according to Pascal's precept, to substitute mentally in that which is defined the Cartesian definition of the word *body* in order to recognize that in Descartes' philosophy motion implies a contradiction? Is it not clear that in order to conceive of motion it is necessary for us to conceive of, in extension, something that is distinct from extension and that stays unaltered although the location changes?

Cartesian matter is incapable of motion; Cartesian motion, in its turn, is incapable of serving to construct a Mechanics.

Descartes only wanted to see, in motion as in the whole of Physics, what a geometer saw. Now, has the geometer a direct and immediate intuition of the state of motion? No; in the spectacle which bodies offer him, he can perceive only one element, the figure; motion is only perceptible mediately to him by the intermediary of the statement following: at the various instants of the duration, the bodies are disposed in such a way as to produce different figures. The geometer can therefore declare that between two given instants two given bodies A and B are displaced with respect to each other; but upon whether to ask of him whether it is A that has stirred, or B, or both of them, there is no need to muse; this question would have no meaning for him; he knows only *relative motion*.

This point did not escape Descartes when he defined motion[2]: *the transport of a part of the matter, or of a body, of the neighbourhood of the body which touches it immediately and which we regard as at rest, in the neighbourhood of other bodies.* He insisted[3], furthermore, for fear that his course of thought might not be clearly understood, that when two bodies, which are continuous, are separated from each other, there is no reason to attribute the motion to the one rather than the other; alone have practice and convenience guided us when we choose one of these bodies as the term immobile.

Now the laws of Mechanics cannot be adapted to this absolutely relative character left to the notion of motion. Their form, universally accepted, implies this consequence: if they are in con-

---

atomes dans la science contemporaine, livre I, chapitre II, §5, (Paris), (1895).

[1] *I.e.*, 'locomotion'. (Translator's note).
[2] Descartes: *Principia Philosophiae*, Pars II, art. XXV.
[3] Descartes: *Principia Philosophiae*, Pars II, art. XXIX, XXX.

formity with the various natural motions when one of the bodies which form the world is regarded as fixed, they cease to accord with these motions when the fixity is attributed to another body. The motions of the stars, for example, agree with a certain Celestial Mechanics when one assigns the fixity to the stars; they violate this same Mechanics when one assumes the Earth to be immobile. Each mechanical explication of the world assumes that the motions are referred to a particular fixed body[1], when one changes the fixed body taken for reference one is obliged to change the form of the Mechanics.

This strange character of the laws of Mechanics explodes in the very law of inertia: A material point, removed in the extreme from every body, moves in a rectilinear and uniform motion. Let us assume that this law is satisfied when one refers the motion of the material point to a certain body regarded as fixed; let us change our frame of reference; let us now attribute the fixity to a new body which was animated, a little while ago, with respect to the first, by an arbitrary motion; our isolated point mass is going to describe the trajectory we shall desire according to the law it may please us to impose upon it[2].

Therefore when Descartes admitted[3] the inertia principle, he himself forgot under what conditions an explication is acceptable in its Physics.

---

[1] More exactly, that the motions are referred to a certain body or to another body whose motion relative to the first reduces to an uniform translation.

[2] A complete historical study of this question will be found in the work: Heinrich Streintz: *Die physikalische Grundlagen der Mechanik*, (Leipzig), (1883).

[3] Descartes: *Principia Philosophiae*, Pars II, art. XXXVII.

CHAPTER III

ATOMISTIC MECHANICS

"All that", said Leibniz, "leads us to recognize that there is in Nature something other than that which is purely geometric, that is to say other than bare extension and its alteration".

Physics was therefore forced to abandon the form of mechanical explication, ideally simple, imposed upon it by Descartes; it was compelled to put into its theories something other than notions available to a geometer, something other than pure extension and purely relative motion; after having dashed imprudently to the conquest of an indefensible position it found itself obliged to shuffle off in retreat.

But in this retiring movement it only retreated step by step; it did not abandon a scrap of territory without having disputed it energetically; driven back from Cartesianism, it billeted itself first of all in the position it occupied when Descartes led it forward, in the atomistic doctrine that Gassendi had borrowed from Empedocles, Epicurus and Lucretius, and which he had renovated. When Huyghens speaks of the "true Philsophie within which one conceives of the cause of all natural effects in terms of Mechanics", it is of the corpuscular theory that he intends to speak.

Certain parts of space remain as pure extension; they form the *void*; others, in opposition to this, are occupied by a material substance; these latter consist in very small volumes, separated from one another by the void; each of the *small bodies* this scattered in the void has a constant geometric form and unchanging dimensions; its *hardness* protects it from any deformation, against any penetration, against any rupture; it is physically indivisible, and from that earns the name *atom*.

In the void each atom moves with a uniform and rectilinear motion until it comes into contact with another atom; at this moment an *impact* is produced; each of the two atoms resumes its course with a different motion from that it had before the impact; the distribution of velocities before the impact and of the *mass* of each of the two atoms that have collided, as each atom has an unchanging mass.

By what law is expressed this dependence between the motions of the atoms before the impact and their motions afterwards? Experience cannot lead us to recognise this law; each of the bodies, amongst which the effects of the impact are observed, is the deformable union of an immense number of atoms. Hence it is necessary, in order to discover it, to have recourse to hypothesis, to invoke arguments which are imposed not without contest. From that there arose long and passionate debates amongst the atomists.

Extolled by Huyghens, Atomistic Physics was to persist throughout the whole of the eighteenth century, in spite of the resounding success won by Newtonian Physics; Daniel Bernoulli devised an explication, that has remained classical, of the expansive force of gases; in Switzerland, around Bernoulli there became grouped a small but brilliant school of geometers which remained faithful to the principles of Epicurean Philosophy; one, even, of these geometers, Lesage, took up the attempt of Fatio de Duilliers and endeavoured to explain the Newtonian laws of universal attraction by atomistic methods.

In order to explain the effects that corporeal nature manifests, the atomists did not only call to their aid purely geometric arguments, they again invoked the *hardness* of atoms, and more than one Physicist entertained that—seeing in this intervention a return to the virtues and qualities of the School. "One thing which causes me difficulty", wrote Denis Papin to Huyghens[1], "is you say that you believe that perfect hardness is of the essence of bodies: it seems to me that there one assumes an inherent quality which removes us from the principles of mathematics or mechanics: for, in the end, an atom, as small as one may wish to take, and however composed of parts really distinct from each other; the oriental half is really distinct from the occidental half, so that if I give a blow to the oriental half only, to push it towards the middle, there is no mechanical reason that obliges me to believe that the occidental part will also travel from the same side; thus, to me it seems that to hold absolutely to the Principles of Mechanics, it is necessary to believe that the matter itself has no parts, and that the hardness which is evidenced in certain bodied only comes from the movement of the surrounding liquids which press the less agitated parts towards each other".

At the same period, Leibniz[2] and Malebranche[3] continued, for analogous reasons, to keep a Physics as close as possible to Cartesian Physics; according to these two great philosophers, a ho-

---

[1] D. Papin à Christiaan Huyghens, 18 juin 1690, *Oeuvres complètes de Christiaan Huyghens*, vol. IX, p. 429.

[2] Leibniz: *Theoria motus concreti, seu Hypothesis nova*, (Moguntioe), (1671).

[3] Malebranche: Réflexions sur la lumière et les couleurs, *Histoire de l'Académie Royale des Sciences, annee 1699, Memoires*, p. 22.

mogeneous matter, infinitely divisible, fluid, incompressible, filled space; alone, the whirling motions distinguished the various parts; the pressures generated by these whirling motions explained the apparent hardness of these parts and the actions that they seem to exercise mutually upon each other.

Thus, the seventeenth century was already close to ending that which some great minds still were struggling to pursue, the Cartesian method, of accepting into their Physics only figure and motion, still ferretting out the qualities of this School to their last refuge, the hardness of Epicurean atoms. But at that very moment a Physics was welling upwards that admitted an idea into its arguments radically different from Geometry, the idea of *force*; this Physics was that of Newton.

CHAPTER IV

NEWTONIAN MECHANICS

We do not intend to give a detailed exposition here of the successive developments of this doctrine; however, who is there who does not know the principal phases of this triumphal march?

In 1687 there appeared the *Philosophiae naturalis Principia mathematica*; in the first two books of this immortal work, from the fundamental axioms of the new Mechanics, stated with a remarkable nicety, their consequences unfold through geometric deductions whose elegance has been admired over the centuries; in the third book of this treatise, attraction, under the inverse square law for distance, allows us to analyse, with a precision hitherto unknown, the motions of the planets, of the satellites and of the waters of the sea. In a celebrated question, added to his *Optics*, Newton guessed that Electricity and Magnetism followed laws analogous to those that control celestial bodies; he imagined a molecular attraction that explained capillary phenomena and chemical reactions. These outlines of actions at very small distances were transformed into a precise doctrine by the investigations of Freind, Keill and Clairaut at the same time that all the great geometers became consumed with contributing to Celestial Mechanics upon the foundation of universal gravitation.

Without detailing the history of this development, we shall take immediately Newtonian Physics in the form it had reached in its full blossom, in the form that Boscovich[1] recorded with such rigour and clarity.

In an empty space there are material objects, each of which is reduced to a point, deprived of any extension, but endowed with an unchanging mass. Each of these points is subject to forces, the resultant of which is obtained by the classical parallelogram law. At each instant the resultant of the forces which

---

[1] Boscovich: *Theoria Philosophiae naturalis redacta ad unam legem virium in Natura existentium*, (Vienna), (1758); (Venice), (1763).

attract a material point is directly opposed to the acceleration of the motion of this point. Between the magnitude of force and the magnitude of the acceleration there exists an invariable ratio that is exactly the mass of the movable point.

The action between two points is directed along the line which joins them; it is proportional to the product of their masses, it varies with the distance that separates them.

When two points are separated by a distance so small that it escapes entirely from the grip of our senses and from the detection of our instruments, the function of this distance, upon which their mutual action depends, has a form that is unknown to us and which may be complicated; this form may change with the chemical nature of the two material points; the action that it represents may be an attraction when the mutual distance has certain values and a repulsion when this distance has other values.

Against this, when the two points are separated by an apprehendable distance, their mutual action becomes independent of their chemical nature; it is always attractive; it varies simply by the inverse square law for the mutual distance.

In the latter form the mutual action becomes the *gravitational attraction* which gives account of the falling of bodies to the surface of the Earth, the progress of the Moon, of the planets, of satellites and of comets, of the ebb and flow of the sea. In the first form the mutual action takes the name of *cohesion* when it is exercised between two point masses of the same nature, and of *affinity* when it is exercised between two points chemically different; cohesion explains the properties of solids, liquids and gases; it concords with affinity for determining and regulating chemical decompositions.

In their essential form, such are the principles upon which the mechanical explication of all physical phenomena rests; such is the general plan of the theories that Laplace's School were to carry to the highest degree of perfection.

"Laplace", said Fourier[1], "was born to perfect everything, to deepen everything, to push back all frontiers, to solve what would have been believed to be insoluble. He would have attained Heaven's science if that science were attainable".

The ceaseless object of attention for all the great geometers of the eighteenth century, for MacLaurin, Clairaut, d'Alembert, Euler and Lagrange, Celestial Mechanics based upon universal gravitation had already had an ample development. Laplace "formed the project of devoting his efforts to this sublime science. He meditated profoundly upon its glorious design; he spent all his life to accomplish this, with a perseverance for which the history of science can offer perhaps no example ... . There is no point of Physical Astronomy that did not become, for him the subject of a study and of a profound discussion; he has subjected

---

[1] *Éloge historique de M. le Marquis de Laplace*, delivered at the public meeting of the Académie Royale des Sciences, 15th June 1829, by M. le baron Fourier.

to calculation the majority of the physical conditions that his predecessors have omitted."

The conquests of Laplace in Celestial Mechanics do not form the entire domain of this powerful genius. "He was almost as great a physicist as a geometer." In all branches of Physical Mechanics he pushed forward the consequences of the Newtonian hypothesis.

Newton had already regarded attraction at small distances as necessary to account for the shape of liquids in very narrow vessels, and he had encouraged Hawksbee to check experimentally the consequences of his insights; Jurin had pursued the application of these views to the rising of water in very slender tubes and Clairaut had set this problem in terms of the exact principles of General Hydrostatics, which he had discovered; a fortunate inference, assuming the analogy of the end surface of a liquid with a tensioned elastic membrane, had led Segner to the equation for the capillary surface and Young to the expression for the angle of contact. But what a vast distance between these various attempts and the complete detailed theory that Laplace gave! This theory, established be extremely elegant geometric methods, rich in precise consequences, minutely controlled by the agreement of its consequences with Gay-Lussac's experiments, can be regarded as the theoretical model for a physical explication conceived in terms of the doctrines of Newton and Boscovich.

In addition, these doctrines, cultivated by Laplace and his disciples, gave many other proofs of their fertility.

Newton had produced the hypothesis that light is formed of very small projectiles moving with an extreme velocity; that the corpuscles of matter exercise upon these projectiles attractions which become very strong if the points reacting are very close. Upon this hypothesis Laplace built his *Optique*; he took it up to the point of giving an account of the double refraction laws of Icelandic spar, the discovery of which, due to Huyghens, had been the greatest work of that great atomistic physicist.

A few years later, Corpuscular Optics, ruined by the prodigious discoveries of Young and Fresnel, was to yield place to Wave Optics; but the same principle of Newtonian explications was not to be shaken off; not before long its fertility was to receive a new increase; it was to this principle, in fact, that Fresnel appealed for support for the laws of the elasticity of the aether; from that it attracted the vigorous attention of geometers of the general theory of the elasticity of solids; and large borrowings from Newton's hypothesis and Laplace's methods were to permit Poisson and Cauchy to take up the work of Navier again, and to build upon this theory.

After the calorimetric investigations of Black and Crawford, for half a century heat lost the character of a motion that had been attributed to it since Descartes; by a return to the hypothesis of Gassendi, it became a fluid, *caloric*; the material points which compose this fluid repel each other, whereas the matter of bodies attracts them. Whilst Lavoisier, then Berthollet, sought to explain by these assumptions the laws of melting, volatilisation, dissolving, chemical reactions, Laplace, assisted by Poisson, aided by the experiments of Desormes and Clément, Delaroche and Bérard, Gay-Lussac, Welter, constructed the explication of the expansion of gases and of the calorific phenomena that accompany them; he represented the velocity of the propagation of sound in air by an exact formula; he set up the foundations of a theory of heat of which several equations were to survive the hypothesis that gave rise to them.

It was again the advice of Laplace that led Poisson to treat the attractive and repulsive actions of electric fluid according to the rules of Newtonian Physics.

Poisson thus discovered the laws according to which electricity is distributed over the surface of a conducting body, then by an extension of the same analysis he gave a detailed theory of the magnetisation of soft iron.

Finally, from 1822 to 1826, Ampère constructed his immortal *Théorie mathématique des phénomènes électrodynamiques*; in that, Newtonian Physics conquered a new empire by submitting electrodynamic and electromagnetic forces to the rules of this Physics.

CHAPTER V

FORCE AND HIDDEN PROPERTIES

Newtonian Philosophy, which was beginning to show its fecundity, was not greeted without some mistrust. The new Physics was successfully solving the problems of Celestial Mechanics which for a century had attracted the efforts of atomistic or Cartesian philosophy; and so it was offending the self-respect of more than one geometer; the friends of Newton did not take to conflict to avoid these skirmishes; they did not even wait for the book *Principia* to be printed to "make it understood that ever since the meditations of its author, all of Physics has been quite changed[1]". But another cause, and one more avowable, began to provoke the hostility of these men so attached to explaining everything by mechanical arguments; the mutual attraction of the various parts of matter too closely resembled the hidden properties that the Scholastics, the Cartesians and the Atomists had pursued with neither respite nor mercy, in order that the latter might in no way be shocked by the form of this hypothesis.

"I would wish, Sir," wrote Fatio de Duilliers to Huyghens[2], "that the author had consulted you about this principle of attraction which he assumes between celestial bodies!."

— "I desire to see the book of Mr. Newton", replied Huyghens[3]. "I like it that he is not a Cartesian, provided that he does not make assumptions like that of attraction." Leibniz, for his part, after having read Newton's book, wrote to Huyghens[4]: "I do not

___

[1] Fatio de Duilliers à Christiaan Huyghens, 24 juin 1687, *Oeuvres complètes d'Huyghens*, vol. IX, p. 163.
[2] Fatio de Duilliers à Christiaan Huyghens, 24 juin 1687, *Oeuvres complètes d'Huyghens*, vol. IX, p. 169.
[3] Christiaan Huyghens à Fatio de Duilliers, 11 juillet 1687, *Oeuvres complètes d'Huyghens*, vol. IX, p. 190.
[4] Leibniz à Christiaan Huyghens, octobre 1690, *Oeuvres complètes d'Huyghens*, vol. IX, p. 523.

understand how he conceives of heaviness or attraction. It seems that according to him this is only a certain incorporeal and inexplicable property." And Huyghens replied to him[1]: "As to this which is the cause Mr. Newton gives for the ebb of tides, I find I am not at all content, neither with all his other theories which he bases on his principle of attraction, which to me seems absurd."

This repugnance that the hypothesis of a mutual attraction between the various parts of matter was having to meet from the minds hostile to hidden properties, to the sympathies and antipathies, had certainly been foreseen by Newton; also, in ending the book *Principia*, had not he taken care to present this attraction as a last explication, as a property irreducible to figure and motion; he left one vaguely to foresee the possibility of such a reduction, in the investigation of which he had himself made some attempts and which Fatio de Duilliers endeavoured to obtain; but he gave it to be understood that the efforts to this end were largely futile for discussions on hidden causes.

"Up to now", he said[2], "I have rendered account of the phenomena presented to us by the heavens and the sea by means of the force of gravity; but to this gravity I have not yet assigned a cause ... . I have not been able, to this day, to show from phenomena the reason for the existence of the properties of gravity, and I make no hypotheses. In fact, all that which cannot be deduced from phenomena must be named *hypothesis*; and hypotheses, whether they be physical or metaphysical, whether they invoke hidden properties or some mechanism, have no place in *experimental philosophy*."

The thought this celebrated passage expresses is made with even more nicety, if that be possible, in these lines from the *Optics*[3]: "To be explaining each property of things by endowing them with a specific hidden quality by means of which there would be generated and produced the effects that are manifested to us, this is to explain nothing at all. But to draw from some phenomena two or three general principles of motion; further to explain the particularities of the actions of the bodies by means of these clear principles, this is, in Philosophy, truly to make great progress, even when the causes of these principles might not be discovered; this is why I do not hesitate to propose the Principles of Motion, quite leaving to one side the investigation of the causes."

The preface inserted by Roger Cotes at the front of the second edition of the *Principia* accentuated the opposition between Newton's philsophy and the methods clear to the Cartesians and Atomists; Cotes there scoffs at the hypothetical explications of these physicists, at the assurance with which they attributed to small parts

---

[1] Huyghens à Leibniz, 18 novembre 1690, *Oeuvres complètes d'Huyghens*, vol. IX, p. 528.
[2] Newton: *Philosophiae naturalis Principia mathematica*; Scholium generale.
[3] Newton: *Optice*, Quaestio XXXI.

of bodies the dimensions and shapes which agree with their arguments, their unobservable fluids that had to penetrate all substances through invisible pores; despite their scrupulous correctness in following the laws of Mechanics, they took only as foundations these misleading conjectures; "the fable that they hold for us is graceful and pretty, but it is only a fable."

After that, however did they come to accuse gravity of being a hidden cause! The answer is easy. Which are the true hidden causes, those that experiment test with complete clarity or those of which the existence is only a fiction? The force whose celestial motions describe all its characters, or the vortices of some subtle matter which escapes every verification?

Will they say that gravity is hidden because the cause of gravity is concealed and has not yet been discovered? But to go upwards from cause to cause one would have to arrive at the simplest causes, and from there it would no longer be possible to give a mechanical explication. Would they be called hidden and would they be ejected from Physics? Physics would then have disappeared completely.

One would remain in no doubt about the profound thinking of Roger Cotes; for him gravity was an inherent property of matter, a first and irreducible quality of corporeal substance.

Leibniz, who in his youth was so strongly attached to the purely geometric explications of the Cartesians, was led himself, also, to turn to admitting in Mechanics a heterogeneous element in extension and motion; still more audacious than Roger Cotes he did not hesitate to assimilate explicitly this element into the substantial forms that the Scholastics invoked:

"Although I am persuaded that everything happens mechanically in coporeal Nature", he wrote[1], "I do not also give up believing that the very principles of Mechanics, that is to say the first laws of motion, have a more sublime origin than that which pure mathematics can furnish... . One perceives that some superior or metaphysical notion has to be added, to wit substance, action and force; and these notions show that everything that suffers must act reciprocally, and that everything that acts must suffer some reaction... . I still accept, naturally, that every body is extended and that there is no extension without a body; nevertheless one must not confuse or identify the notions of place, space nor of pure extension with the notion of substance, which, in addition to extension also include resistance, that is to say action and rest.

"I had already gone into the territory of the Scholastics", he wrote elsewhere[2], "when mathematics and the modern authors had made me depart from thence, whilst still quite young. Their beautiful ways of explaining Nature mechanically charmed me and I scorned, with reason, the methods of those who used only forms and means of which nothing could be apprehended. But since then,

---

[1] Leibniz: *Oeuvres*, (édition Gerhardt), vol. IV, p. 464.
[2] Leibniz: *loc. cit.*, p. 478.

having tried to deepen the very principles of mechanics to give reasons for the laws of Nature that experiment had brought to knowledge, I realised that the single consideration of an *extended mass* was not sufficient, and that it was still necessary to employ the notion of *force*, which is extremely intelligible, although it does resort to metaphysics.

"And by *force* or *power* I do not understand[1] the ability or the simple property which is only a subsequent possibility for acting and which, being dead itself, produces no action without being excited from outside; but I understand by it something between the ability and the action which surrounds an effort, an act, an entelecheia, for force passes from itself into action, insomuch as nothing impedes it.

This passage, and many another that it would take too long to cite, proves to us that Leibniz's ideas recaptured a direct contact with the antiquarian Peripatetic Physics. "I know", he said[2], "that I advance a considerable paradox in pretending a rehabilitation, in some manner, of the ancient philosophy, and in recalling *postliminio* substantial forms, which had almost been banished; but perhaps one will not lightly condemn me when one knows that I have thought a lot about the modern philosophy, that I have given plenty of time to physical experiments and to proofs in geometry, and that I have long been persuaded of the vanity of things, that I have at last been obliged to return to these things, despite myself and as if perforce, after having myself made some investigations which have made me recognise that our moderns do not do enough justice to St. Thomas and other great men of this time, and that there are in the thought of the scholastic philosophers and theologians more of solidity than has been imagined, provided that they are used appropriately and in their place. I am even persuaded that if some exact and meditative mind were to take the trouble to clarify and digest their thought in the manner of the analytic geometers, he would find a treasury of very important truths and completely demonstrable.

Not that one has to approve or even imitate these ridiculous methods of Physics which had so powerfully discredited Scholasticism: "I still accept[2] that the consideration of these forms gives nothing to the detail of Physics nor has to be employed for the explication of phenomena in the particular. It is in what our Scholastics have missed, and physicians of the time followed in their footsteps, believing to give reason to the properties of bodies by making mention of forms and qualities, without taking care to examine the manner of their operation, as if one were content to say that a clock has the horological quality originating from its form, without considering what is consists of.

Far from imitating this Physics, which was believed to have given an explication when it has only created a name, one would have to push the analysis of natural effects, imitating Descartes

---

[1] Leibniz: *loc. cit.*, p. 471.
[2] Leibniz: *loc. cit.*, p. 434.

and Huyghens, until they were reduced to the simplest phenomena; but when one would have reached these first properties of bodies, which explain all others, one would find that they consist "not solely in extension[1], that is to say in magnitude, figure and motion, but that it would be necessary therein to recognise something which would have some relation with souls, and that one commonly calls substantial form", or *force*, as Leibniz says in many a place.

Leibniz had started from a system in which he had rejected attraction, for to him it seemed to be "a certain incorporeal and inexplicable property"; his reflections upon the foundation of Mechanics had led him to share, on the nature of this property, the opinion of the immediate disciples of Newton and to bring vividly to light the analogy between this opinion and the peripatetic doctrines.

Amongst the successors of Newton the most diverse opinions came to be admitted, touching upon the nature of attraction.

Some under the influence of Bernoulli continued to "simulate hypothesis" so as to reduce all the effects of corporeal Nature to the only reasons received by the atomists; amongst those, Lesage, renewing the attempt of Fatio de Duilliers, tried to explain gravitation by the impact of *ultra-mondain corpuscles* upon the material molecules.

Some others had no scruples about invoking in their arguments forces exercised or suffered by the various material points that formed the bodies; but they copied the prudent reservation that Newton had put in his *Principia*; they undertook in no way to decide whether attractions had to be regarded as irreducible properties of matter, or, contrary to that, as the effects of suitably imagined motions. It is amongst the latter that we must place Laplace. The principle of universal attration, he said[2], "is it a primordial law of Nature? Is it only a general effect of an unknown cause? Here, the present ignorance of the intimate properties of matter halts us, removes from us every hope of answering these questions in a satisfactory manner". — "The principle of universal heaviness", he further said[3], "is it a primordial law of Nature, or is it a general effect with an unknown cause? ... Newton, more cautious than several of his disciples, made no pronouncement on these questions, in which our present ignorance of the intimate properties of matter does not allow us to anser in a satisfactory manner."

And finally, yet other, pursuing to the end the ideas of Leibniz, did not hesitate to see in *force* a notion irreducible to extension and motion, a first and essential property of material substance. Amongst these the first place belongs to Boscovich[4],

---

[1] Leibniz: *loc. cit.*, p. 434
[2] Laplace: *Exposition du Système du Monde*, livre IV, chapitre XVII.
[3] Laplace: *ibid.*, livre V, chapitre V.
[4] Boscovich: *Theoria philosophiae naturalis redacta ad unam*

who proclaimed himself to be at the same time both a disciple of Leibniz and of Newton, and who gave Newtonian Physics a form admirable in unity and rigour.

---

legem virium in Natura existentium, (Vienna), (1758); (Venice), (1763).

CHAPTER VI

THE PRINCIPLE OF VIRTUAL VELOCITIES
AND LAGRANGE'S STATICS

In treating attractions and, the converse, repulsions of material points as realities irreducible to figure and motion; in considering them, contrary to this, as the effects of motions hidden yet from our investigations, this is nothing less than the physicist can and must invoke in his arguments, not just figures and explicit motions, but, further, *forces*, actually heterogeneous to the notions of Geometry and Kinematics. From this the words: *To explain a physical phenomenon* take on a meaning quite different from that the Cartesian and atomistic philosophers attributed to them; the explication that is halted at *force*, taken as an element that is actually or provisionally simple, has an analogy with the scholastic explication in terms of hidden qualities and properties.

According to Newton, just as according to Leibniz, what must essentially distinguish the new Physics from the Physics of the Old School is the *generality of its principles*; it no longer has to render an account of each phenomenon by creating for the purpose a new and special cause; it must unravel all the detail of the observed facts in coporeal nature by invoking a minimum number of principles that are as comprehensive as possible.

Certainly the Physics of which Newton had drawn the plan and proposed the bases, of which Boscovich had analysed the complete structure, was already admirable for the simplicity and fullness of its principles; however, besides the fundamental hypothesis that in the world there are: time extension, mass and force, did not this Physics admit other assumptions that could be eliminated? In place of reducing mass to a set of points without any extension and isolated from each other, could one not in it conceive of extended bodies, variable in their figure, capable of touching? In place of regarding all forces as attractions and, conversely, repulsions, functions only of the distances separating them from the points they attract, could one not leave them entirely undetermined, ascribing only to each action an equal and directly opposed reaction? Would not, then, the principles of Mechanics be thus led to the highest degree of generality of which one could conceive?

To this construction of Rational Mechanics contributed the greatest geometers of the eighteenth century; Daniel Bernoulli, d'Alembert, Euler, to cite only the most illustrious, fixed their name to some part of the edifice; but to Lagrange is owed its complete achievement.

"Lagrange", said Fourier[1], "was born to invent and enlarge all the sciences of calculation. In whatever position fortune might have set him, be it shepherd or prince, he would have been a great geometer; he would have become so of necessity, and effortlessly...

"The distinctive feature of his genius consisted in the unity and breadth of his vision. In everything he applied himself to a simple thought, correct and very deep. His principal work, *Mécanique analytique*, could be called Mécanique philosophique, for it reduced all the laws of equilibrium and motion to a single principle; and, what is no less admirable, he submitted them to a single method of calculation of which he, himself, was the inventor."

The first part of *Mécanique analytique* is devoted to *Statics*; it begins with these words:

"By *force* or *power* one understands, in general, the course, whatever it may be, which impresses, or tends to impress, motion upon the body to which it is supposed applied, and that it is also by the quantity of motion impressed, or ready to be impressed, that the force or power may be expressed. In the state of equilibrium the force has no actual effect; it produces only a simple tendency to motion; but one must always measure it by the effect that it would produce if it had not been arrested."

In the margin of the copy of the *Mécanique analytique* which had guided his meditations, Saint-Venant wrote these words: "Thus the author of the *Mécanique Analytique* has no doubt about the existence of forces or special causes of each motion". In fact the passage we have just quoted reproduces the ideas, and almost the very terms, of certain fragments of Leibniz; like Leibniz, Lagrange regarded the notion of force as one of the first notions of Mechanics; if he invoked motion, it was not to explain force, it was only to make it correspond with that idea, transcendent of Geometry, a numerical symbol able to appear in formulae.

First of all, Lagrange occupied himself with proposing the principles of Statics, that is to say to fix the circumstances under which the forces applied to a material system would hold it in equilibrium.

The problem of Statics was eased in Newtonian Physics; as every system reduces to free points, the equilibrium of a system follows from the equilibrium of each point; and each point will

---

[1] *Éloge historique de M. le marquis de Laplace*, delivered at the public meeting of the Académie Royal des Sciences, 15th June 1829, by M. le baron Fourier.

be in equilibrium when it is attracted by forces whose resultant is zero; thus the whole of Statics may be drawn from the single rule of the parallelogram of forces.

The question is delicate in another way when one restores to bodies their extension, their figure, the possibility of sliding or rolling upon each other, of being deformed, indeed.

Already, for discussing the equilibrium of similar systems, at least in very simple cases, Archimedes had proposed the principle of the equilibrium of levers. The long elaboration, which, in the course of modern times, has shaped Mechanics, transformed little by little this ancient rule into a new principle: the *Principle of Virtual Displacements*.

To recover the source of the Principle of Virtual Displacements one has to go back to the Renaissance, to Leonardo da Vinci, to Guido Ubaldi; this is made precise in the writings of Galileo, whose arguments are a commentary upon this formula: "The gain of power which empowers a mechanism implies an equivalent loss of velocity"; in Descartes, who starts from this proposition: "The same force that can raise a weight of one hundred pounds by two feet can raise one of two hundred pounds to a height of one foot"; in Toricelli and Pascal, who admit this principle: "Never is a body moved by its weight except that its centre of gravity descends".

Some little forsaken whilst Huyghens and Newton were creating the science of motion, the Principle of Virtual Velocities, or, better, of Virtual Displacements, was taken up again in a more complete and more general form by John Bernoulli I, who communicated it in 1717 to Varignon; the latter, in his *Nouvelle Mécanique*, gave it numerous applications; but it was kept to Lagrange to discover in it a base large enough upon which to repose the whole of Mechanics[1].

The bodies which compose a material system cannot be tested, no matter what the alteration of form or position; the nature that is attributed to them, which serves to name them, which properly constitutes their definition, excludes certain displacements, certain deformations that it would be contradictory to attribute to them. Is a body solid? Its location may change, but its figure and dimensions must stay unchanged. Are two bodies in contact? They may roll and slide on each other, but without penetrating each other or being deformed. A flexible and inextensible thread may draw all sorts of lines provided that its length does not change. An incompressible fluid can occupy spaces that are the most varied in shape, provided that they all have the same volume. One calls *constraints* these restrictive conditions which follow from the very definition of a mechanical system, and calls *constraint equations* the algebraic equalities by which these conditions are expressed.

If one does not wish to contradict the very definition of a

---

[1] Lagrange: *Mécanique analytique*, Part I, Section II. (We always cite this work in its second edition, the last to which Lagrange laid his hand).

system, one cannot impose all imaginable displacements upon the bodies which compose it, but those alone that are compatible with the constraints; these are those displacements which one calls *virtual displacements*.

Let us impose upon the system we wish to study an infinitesimal displacement; the point of application of each of the forces which attract the system describes an infinitesimal path that can be treated as being rectilinear; let us take the component of the force in the direction of the infinitesimal path and multiply the size of the component by the length of the path; the product so obtained will be called the *work* of the force along the infinitesimal displacement under consideration; if the displacement is virtual, the work will be a *virtual work*.

We are now in a position to enunciate the Fundamental Principle of Statics: *In order that a set of forces hold a material system in equilibrium, it is necessary and sufficient that every virtual infinitesimal displacement imposed upon the system cause the sum of the virtual works performed by the forces to vanish.*

How many new and fertile ideas came to be grouped around this principle in the first part of *Mécanique analytique*! Lagrange expounded it in only a few lines, but its far reaching nature is reaffirmed considerably every day.

It is clear that the virtual work of a set of forces applied to a system that suffers a given displacement simply changes sign if one reverses the sense of all the forces without changing either the magnitude of points of application of each of them. From there let us imagine two sets of forces, mutually different, but which when successively applied to the same material system produces the same work for every virtual displacement. Let us apply them simultaneously after having reversed the sense of the forces of one of them. Every virtual displacement will now generate zero virtual work, so the system will be in equilibrium. Thus any one of our two systems is equilibrated by the other when its forces have their senses reversed. In other words our two sets of forces are exactly equivalent for the material system under study; one can be substituted for the other without changing any of the mechanical properties of the system.

It is therefore not too important to know in detail each of the forces applied to the various bodies of the system, its point of application, its magnitude, its direction; provided that the information given about the set of forces allows the calculation of the work done over a virtual displacement, enough is known; all other additional data are superfluous; information of a different form, but which leads to the same expression for the virtual work, will be identifiable to a mechanician's eyes[1].

Thus it is that for the various forces applied to a solid body one can substitute a certain set of two forces, either a force and a couple, or yet other combinations of forces; all these com-

---

[1] Lagrange: *Mécanique analytique*, (2nd edn), Part I, Section II, n° 14.

binations which appear distinct to a geometer, provide the same work for a virtual displacement of the solid body; a mechanician therefore does not distinguish the one from the other.

It follows from this that by studying the equilibrium and motion of a solid a mechanician will not be able to decide whether the group of forces to which this solid is effectively subject is the one or the other of the combinations; a geometer will think that there is some indetermination of the solution of the problem, but not the mechanician, if he has meditated but ever so little upon the principles of the science he cultivates; he will see clearly that for every set of forces capable of producing the motions that are observed in a system one can substitute an infinity of others which would produce the same motions.

To define the infinitesimal virtual displacement, of a mechanical system, in all its generality, it is not necessary, in the majority of cases, to be given the magnitude and direction of the path traced by each of the material points; it suffices to be given the values of certain suitably chosen infinitesimal quantities; the propositions which fix the nature of the system, when these values are known, will allow the determination of the path described by any such point as may be wished.

Let us assume, for example, that the system studied is a solid body; a well known theorem informs us that this solid will always be able to be taken from one arbitrarily specified position to another arbitrary position by the following method: Having selected three mutually perpendicular lines issuing from a point, one gives a suitable rotation to the solid successively around each of these three lines, and then a suitable translation in the direction of each of these lines. The knowledge of the three rotations and translations will determine the trajectory followed by an arbitrary point of the solid; it will fix the virtual displacement completely.

Let us imagine that every virtual infinitesimal displacement of a certain system is thus fully known when one is given the infinitesimal variations $\delta\alpha, \delta\beta, \ldots$ suffered by certain more or less numerous quantities $\alpha, \beta, \ldots$, these quantities are the *independent variables* which define the system.

The expression for the virtual work performed by the forces applied to the system will take the form[1]: $A\delta\alpha + B\delta\beta + \cdots$.

To know all the effects of the forces which act upon the system, it is necessary and sufficient to know the expression for their virtual work; and to know this expression it is necessary and sufficient to know the quantities $A, B, \ldots$. Thus it is the knowledge of these quantities that is of real importance to a mechanician, and not that of the forces by means of which they are assumed formed. One can give to these quantities the name of *generalised forces*[2].

---

[1] Lagrange: *Mécanique analytique*, (2nd edn), Part I, Section II, n° 12,13.

[2] Lagrange: *ibid.*, n° 9.

The nature of a generalised force A depends upon the nature of the variable α to which it refers, for the product $A\delta\alpha$ must always represent a work. If α and, therefore, $\delta\alpha$ are lengths, A is a *properly called force*, but if α and $\delta\alpha$ are angles, A is a quantity of the same kind as the *moment of a couple*; if α and $\delta\alpha$ are surfaces, A will be called a *surface tension*; if α and $\delta\alpha$ are volumes, A will be the same as a *pressure*.

When the virtual work of the forces applies to a system has been put into the form $A\delta\alpha + B\delta\beta + \cdots$, where the generalised forces $A, B, \ldots$ obey known laws, the equilibrium conditions for the system are obtained immediately in the most general and most simple form; they must annul each of the quantities $A, B, \ldots$.

Analysts known that in general an expression such as $A\delta\alpha + B\delta\beta + \cdots$, where $A, B, \ldots$ depend upon $\alpha, \beta, \ldots$, is not the reduction of a quantity that is entirely known when one knows the values of $\alpha, \beta, \ldots$. But this proposition, generally false, becomes exact in certain particular cases; the work performed under arbitrary virtual displacements is then the reduction suffered, under this displacement, by a certain quantity which for each state of the system takes a specified value[1]. To this quantity Lagrange gave no particular name; nowadays it is called the *potential* of the forces that act upon the system.

The existence of a potential for the forces which act upon a system appears as an exceptional property to a mathematician; but if one assumes that a system is subject only to the reciprocal actions of the material points of the volume elements which form it; if one admits, with Newton, that the reciprocal action of two elements is an attraction or a repulsion; that the magnitude of this action is obtained by multiplying the masses of the two elements by a function of their mutual distance; it is then that the forces admit a potential; the study of a system which admits a potential therefore includes as a special case the study of a system isolated in space and constituted as Newtonian Physics requires. If we are limited to the study of systems whose *internal actions* admit of a potential, it will seem to the geometer that we confine ourselves to an infinitely particular problem; however, this problem will infinitely surpass, in fullness and generality, the problem posed by Newton and his disciples.

The mechanical effect of a set of forces depends uniquely upon the expression for their virtual work; when the forces admit of a potential the work they perform in an arbitrary virtual modification can be calculated, provided that one knows the value of the potential and each of the states of the system; this knowledge then replaces and makes useless knowledge of the forces or of the generalised forces. Thus to fix completely the intrinsic mechanical properties of a set of bodies, it is no longer necessary to detail either the forces which are exercised within this set, or the generalised forces to which they are equivalent; it suffices to indicate how the *internal potential* varies with the

---

[1] Lagrange: *ibid.*, n° 21.

state of the system.

Let us press on further: We may, if we wish, consider in Mechanics only groups of bodies completely isolated in space; for this it suffices us to consider a single set and the particular system that we wish to study, and the bodies whose influence upon this system does not seem negligible to us. We shall then have to do only with the mutual forces acting amongst the various bodies of the same system; these internal forces one assumes to depend upon a potential, the knowledge of which renders useless the knowledge of the same forces. Thus the notion of *force*, after having been founded upon a fuller notion, that of *generalised force*, loses, so to say, its first and irreducible character and appears as a simple derivation of the notion of *potential*[1]; such is the natural consequence of the principles proposed by Lagrange, a consequence that accords fully with the proposed views of Leibniz.

The fecundity of these principles is not yet exhausted. They are going to furnish us with a new notion whose role will be considerable in the debates touching upon Rational Mechanics, the notion of *constraint force*[2].

Let us consider two systems which we shall denote by the figures 1,2. The state of the system 1, taken in isolation, is fixed by the independent variables $\alpha_1, \beta_1, \ldots$; the virtual work of all the forces that act upon it is $A_1 \delta\alpha_1 + B_1 \delta\beta_1 + \cdots$. The state of the system 2, taken in isolation, is fixed by the variables $\alpha_2, \beta_2, \ldots$; the virtual work of all the forces which act upon it is $A_2 \delta\alpha_2 + B_2 \delta\beta_2 + \cdots$.

Now, *without altering any of the forces that actually act upon the systems 1,2*, let us juxtapose these systems so that certain bodies of the first system are in contact with certain bodies of the second system, and let us consider the union of these two systems to form a single system.

Each virtual displacement of the resultant system will impose upon the quantities $\alpha_1, \beta_1, \ldots, \alpha_2, \beta_2, \ldots$ some infinitesimal variations $\delta\alpha_1, \delta\beta_1, \ldots, \delta\alpha_2, \delta\beta_2, \ldots$, and upon the forces acting a work $A_1 \delta\alpha_1 + B_1 \delta\beta_1 + \cdots + A_2 \delta\alpha_2 + B_2 \delta\beta_2 + \cdots$; but, and this is the essential point in these considerations, one will not always obtain a virtual displacement of the resultant system by combining any one virtual displacement of system 1 with any one virtual displacement of system 2; each of these two virtual displacements was conceivable when each of the systems 1 and 2 existed alone; their set may become inconceivable when the systems 1 and 2 are juxtaposed, because it could have the effect of bringing, at the same time, certain bodies of system 1 and certain bodies of system 2

---

[1] In 1743, Clairaut became the first to present the concepts of the potential function and the general vector field. The potential concept was first introduced into hydrodynamics in 1752 by Euler in his velocity potential S: $v = \nabla S$, which gave the potential equation $\nabla^2 S = 0$. (Editor's note).

[2] Lagrange: *Mécanique analytique*, (2nd edn), Part I, Section IV, §1.

VIRTUAL VELOCITIES AND LAGRANGE'S STATICS

to the same place. The juxtaposition of the systems 1,2 therefore imposes new restrictions upon the displacements of each of them; these constraints no longer leave completely arbitrary the infinitesimal values that one may attribute to $\delta\alpha_1, \delta\beta_1, \ldots, \delta\alpha_2, \delta\beta_2, \ldots$ in a virtual displacement; they require that these values satisfy one or more equalities called *constraint equations*.

Worries about mathematical generality in what is written here have to yield place to the desire to express these ideas in the simples and most salient form. So let us assume that the union of the systems 1,2 has given rise to a single constraint equation:

$$a_1 \delta\alpha_1 + b_1 \delta\beta_1 + \cdots + a_2 \delta\alpha_2 + b_2 \delta\beta_2 + \cdots = 0.$$

To find the equilibrium conditions of the system we have not only to express that every set of values attributed to $\delta\alpha_1, \delta\beta_1, \ldots, \delta\alpha_2, \delta\beta_2, \ldots$, annuls the virtual work, but only that this work is zero every time that the constraint condition is satisfied. Algebra then teaches us that one can find a certain factor λ, depending upon the state of the systems 1,2 and some forces which act upon them, by which the preceding problem is reduced to this: To annul, for every set of values of $\delta\alpha_1, \delta\beta_1, \ldots, \delta\alpha_2, \delta\beta_2, \ldots$, the sum of virtual works of the first member of the constraint equation, the latter having first of all been multiplied by λ. Thus the equilibrium conditions of our complex system will be obtained as the following:

$$A_1 + \lambda a_1 = 0, \qquad B_1 + \lambda b_1 = 0, \qquad \ldots,$$

$$A_2 + \lambda a_2 = 0, \qquad B_2 + \lambda b_2 = 0, \qquad \ldots,,$$

Let us take the equations of the first line; these are the ones we would have obtained immediately as the equilibrium conditions of the system 1 if we had treated it by leaving out the constraint that contact with the system 2 contributes to its displacements, and by assuming it to be subject not to the generalised forces $A_1, B_1, \ldots$, but to the generalised forces $A_1 + \lambda a_1, B_1 + \lambda b_1, \ldots$. The equations in the second line suggest to us analogous remarks for the system 2.

From there we see that one can obtain the equations of equilibrium of each of the two systems in two ways, distinct in appearance, but rigorously equivalent.

In the first way one regards each of the two systems as subject to the forces that actually do act upon it, but one takes account of the restrictions that their mutual contact imposes upon the virtual displacements of each of them.

In the second way one treats each of the two systems as if it existed on its own, but to each of the generalised forces, such as A_1, to which it is actually subjected, one adds a *purely fictitious* generalised force λa_1; the form of this *constraint force* depends upon the nature of the constrain condition and upon the expression for the factor λ, which is called the *Lagrange multiplier*[1].

One can characterise briefly the connections that exist between these two methods by saying that the first consists in preserving the constraint conditions and avoiding the introduction of constraint forces, and that the second consists in suppressing the constraint conditions and introducing constraint forces.

To bring to light the foundations of Lagrange's Statics we have considered a system whose state is fixed entirely by a larger or smaller, but limited, number of independent variables; all systems cannot be so defined; this is so of continuous media, which have to be decomposed into an unlimited number of infinitesimal elements contiguous with one another; each of these elements depends upon a limited number of variables. Amongst these continuous systems some, such as filaments or elastic rods, extend in one dimension; others, such as membranes or sheet, spread over two dimensions; yet others, like fluids or elastic solids, have finite extension in all dimensions. The principles whose main features we have just described will be applied to such systems with no need to modify them greatly[2]. Only the expression for the virtual work, instead of being simply a sum of terms, will be represented by a simple, double, or triple integral; it will no less be subject to the rules of the calculus of variations.

From this the laws of equilibrium of filaments and flexible membranes[3] take a singular clarity and generality; but above all it is the study of the equilibrium of liquids which tests the breadth and penetration of Lagrange's methods.

Without doubt, since the time of Archimedes Hydrostatics had made incontestable progress. Galileo, Stevin and Pascal had begun, after much groping about, to discover the exact laws of equilibrium of heavy fluids. The problem of the shape of the planets had caused geometers to analyse fluid bodies acted upon by forces other than simple weight; for the attempts of Huyghens, Newton, Bouguer and MacLaurin, Clairaut had substituted a general and rigorous method; in his little Treatise[4], a masterpiece of clarity and elegance which he published in 1743, he gave the general formulae for the equilibrium of fluids, established the links that exist between Hydrostatics and the theory of total differentials, proved that a fluid cannot be brought into equilibrium by every kind of force, and finally discovered the essential properties of level surfaces; in 1755 Euler had rediscovered Clairaut's results by a different process; this same process was one day to allow

---
[1] Euler discovered this multiplier method first in 1732 and used it subsequently in many extremal problems.
[2] Lagrange: *Mécanique analytique*, (2nd edn), Part I, Section IV, §11.
[3] Lagrange: *Mécanique analytique*, (2nd edn), Part I, Section V, Chapter III.
[4] Clairaut: *Théorie de la figure de la Terre, tirée des principes de l'Hydrostatique*, (Paris), (1743).

Cauchy to establish the laws upon which depend the pressure in the midst of an arbitrary body.

Nevertheless, despite this constant progress, everything was not clear and rigorous in the theory of the equilibrium of fluids; the nature of hydrostatic pressure remained quite obscure; it was admitted that this pressure existed, that it was always normal to the surface element to which it referred, that its magnitude did not vary when this surface element turned about one of its points; but of these propositions one had not proof, and even no precise definition of the pressure.

Lagrange[1] obtained the set of all these propositions by using his general method: hydrostatic pressure was introduced into his arguments as one of these multipliers the Calculus of Variations uses to rid itself of the constraint conditions; from there the definition of this pressure is linked intimately to the notion of constraint force.

Let us persist a little with this definition, for it has given rise to serious debates, of which we shall speak later.

Let us imagine that a surface S divides a fluid in equilibrium into two parts A,B; when the fluid experiences a virtual displacement, the two parts A,B do not penetrate each other; this displacement would be known, therefore, not to result in a completely arbitrary displacement of the part A, taken in isolation, coupled with a completely arbitrary displacement of the part B taken in isolation; contact between the two parts constitutes a *constraint* for each of them.

Let us allow the part A to keep its shape and position; let us remove the part B, *but leave unaltered all the faces that actually act upon* A; if amongst these forces there are some that emanate from B, let us imagine that they are replaced by other forces which are exactly equal, but issuing from certain bodies not contiguous with A.

Freed from the obstacle that opposed its contact with the part B, in general the fluid A will no longer be in equilibrium; Lagrange's method proves that it will be brought back into equilibrium if one applies to each element dS of the surface S a force normal to the surface or penetrating the interior of the region A, and of magnitude ΠdS. The factor Π remains unchanged if the element dS turns around one of its points; it represents the hydrostatic pressure at this point.

Hence when the two parts A,B of the fluid are in contact, the hydrostatic pressure acts in fact neither on one nor the other; but if in thought one of them is removed in order to treat the other as if it existed on its own, one must apply hydrostatic pressure to it so as to replace the obstacle that had opposed its motion.

---

[1] Lagrange: *Mécanique analytique*, (2nd edn), Part I, Section VII, §11.

CHAPTER VII

D'ALEMBERT'S PRINCIPLE

AND LAGRANGE'S DYNAMICS

The investigations of Galileo into acceleration during the fall of heavy bodies, of Huyghens into centrifugal force in circular motion, led Newton to postulate the law of the motion assumed by a material point under the action of a force arbitrarily specified. On the one hand, let us consider the line which represents this force, on the other hand let us consider a line in the same direction as the acceleration and equal in length to the product of the latter with the mass of the point; in all circumstances these two lines have the same direction and the same length.

This principle suffices to reduce completely to equations the problem of Dynamics if, in accordance with the rules of Newtonian Philosophy, one reduces all the bodies to material points exercising upon each other attractions or repulsions. Contrary to this, it becomes insufficient if one wishes, as did Lagrange, to treat finite dimensional bodies contiguous with each other and subject to varied constraints. It is then necessary to make use of a Principle of which the previous one is only an extremely particular case.

The invention of this general principle, devoted to putting into equations all the problems of Dynamics, was the object of long and powerful efforts, of which Lagrange has retraced for us the history[1]; these efforts resulted in the discovery of d'Alembert's Principle.

We have considered at an instant in the motion of a material point, the line which is pointed in the same direction as its acceleration and for measure has the product of the mass with the acceleration; we have seen that this line has incessantly been identified with the force.

Without changing the length or the direction of this line, let

---

[1] Lagrange: *Mécanique analytique*, (2nd edn), Part II, Section 1.

us reverse its sense; the new line will be assumed to represent a force which we shall call the *inertial force*. We shall then be able to enunciate the fundamental principle of Dynamics of the material point by saying that the force which actually acts upon this point is, at each instant, equal and directly opposed to the inertial force; or again that the force really acting and the fictitious inertial force form at each instant a set of forces capable of maintaining the material point in equilibrium.

It suffices to generalise this latter enunciation in order to obtain d'Alembert's Principle.

Let us take an arbitrary mechanical system forced of material points or of continuous bodies, and let us divide it into elementary volumes, at each point or at each element we can imagine that one applies an inertial force; directed in a sense contrary to the acceleration of the point or of the element, it will have for measure the product of this acceleration with the mass of this point or of this element. *At each instant the set of forces which actually act on the system and of fictitious inertial forces will be capable of keeping the system in equilibrium in the same state that it presents at this instant.*

This *postulate* — one could not know how to give it another name, despite the arguments, visibly insufficient, by d'Alembert and others after him to try to justify it — was thought up to treat the resistance of fluids in a rational manner[1]. After having shown the utility in Dynamics of systems formed of arrangements of solid bodies[2], d'Alembert applied it once again to the motions of fluids[3]; thus it was that he reached the equations of Hydrodynamics, from which Euler was soon to draw plenty of admirable consequences.

D'Alembert's Principle reduced the reduction to equations of any problem of Dynamics to the reduction to equations of a problem in Statics; now, to treat the latter problem, Lagrange gave a general formula drawn from the Principle of Virtual Velocities; this formula is now going to be extended to produce[4] the "general formula of Dynamics for the motion of an arbitrary system of bodies". This formula will express that every virtual displacement imposed upon the system, starting from the state that it presents at an arbitrary instant, makes the sum of the works performed by the real forces and of the fictitious inertial forces take the value zero.

All the essential ideas introduced by Lagrange in the study of Statics were thus taken over into the study of Dynamics, and their fertility was thereby increased immensely.

---

[1] d'Alembert: *Essai d'une nouvelle théorie de la résistance des fluides*, (Paris), (1742), Chapter I, Proposition 1.
[2] d'Alembert: *Traité de Dynamique*, (Paris), (1743).
[3] d'Alembert: *Traité de l'équilibre et du Mouvement des fluides pour servir de suite au Traité de Dynamique*, (Paris), (1744).
[4] Lagrange: *Mécanique analytique*, (2nd edn), Part II, Section II, n° 5.

The reduction of equations of the problem of Statics takes the simplest possible form when the most general virtual definition of the system is defined by the infinitesimal variations $\delta\alpha, \delta\beta$ of the independent variables $\alpha, \beta, \ldots$. The virtual work of the forces actually acting then takes the form $A\delta\alpha + B\delta\beta + \cdots$; the *generalised forces* $A, B, \ldots$ depend upon the variables $\alpha, \beta, \ldots$.

Analogously, the virtual work performed by the inertial forces can be put into the form $J_\alpha \delta\alpha + J_\beta \delta\beta + \cdots$; the quantities $J_\alpha, J_\beta, \ldots$, which it is natural to name *generalised inertial forces*, depend upon the variables $\alpha, \beta, \ldots$, upon their first derivatives with respect to time, which will be called *generalised velocities*, and upon their second derivatives with respect to time, which will be called *generalised accelerations*.

Lagrange has given[1], in addition, a rule of extreme elegance for forming these quantities $J_\alpha, J_\beta, \ldots$; this rule brings in a quantity which is going to play an essential rôle in Dynamics: the *live[2] force of the system*. This live force is obtained in the following way: Taking each of the points or volume elements that compose the system, one multiplies half of its mass m by the square of its velocity v and one forms the sum of the products so obtained; this sum $\frac{1}{2}(mv^2 + m'v'^2 + \cdots)$ is the live force.

The live force can be expressed by means of the independent variables $\alpha, \beta, \ldots$ and the generalised velocities; the expression for the live force in terms of these elements possesses this double property of being homogeneous and of second degree in the generalised velocities, and of being positive however little the systme may be in motion. Making use of the first of these two properties, Lagrange instituted a regular calculation which, with this expression for the live force, yields the generalised inertial forces.

The fundamental formula of Dynamics requires that the sum of the two quantities $A\delta\alpha + B\delta\beta + \cdots$ and $J_\alpha \delta\alpha + J_\beta \delta\beta + \cdots$ be equal to zero for all the virtual displacements imposed upon the system, or, in other words, that at each instant one has:

$$A + J_\alpha = 0, \qquad B + J_\beta = 0, \qquad \ldots\ldots .$$

Thus are obtained, in the simplest and most easily handled form, the equations which govern the motion of the system.

These *Lagrange's Equations* are in number the same as the independent variables $\alpha, \beta, \ldots$; they connect between them not only these variables, but also their first and second derivatives with respect to time; hence they form what geometers call a *second order system of differential equations*.

One does not have to depend upon us to expound, even summarily, the works to which these equations have given station since

---

[1] Lagrange: *Mécanique analytique*, (2nd edn), Part II, Section II, n°7.
[2] Leibniz coined the term 'live force', which was originally double the kinetic energy; the potential energy of forces Leibniz called the 'dead force'. (Editor's note).

the epoch of Poisson, Cauchy, Pfaff, Hamilton, Jacobi, up to our time, illustrated by the investigations of Henri Poincaré, Painlevé and Hadamard; it is the very history of second order differential equations that we shall be led to write; let us say only that one of the principal and analytic facts brough into evidence by this history would be the extreme importance of the notion of *potential*, introduced by Lagrange.

This importance, moreover, had already exploded into view by the rapid examination of some questions of Dynamics taken amongst the most essential ones; these questions are all connected with the *live force equation*[1].

The basis of this equation is found in the following quite simple remark: When a system moves during a certain lapse of time, the inertial forces perform a work that is precisely equal to the decrease in the live force during the same time. From there it suffices to use the fundamental formula of Dynamics in treating as a virtual displacement each of the elements of the real motion to obtain the following proposition:

The work performed during a certain lapse of time by the actual forces which act upon a system is equal to the increase gained at the same time by the live force of the system.

And so we have precisely, and reduced to the very principles of Mechanics, the celebrated law of the live force, recognised in the first place by Leibniz.

When the forces to which the system is subject admit a potential, this law takes a quite remarkable form; in fact, in this case the work performed during a certain lapse of time is equal to the decrease of the potential during this time; this diminution of the potential is therefore equal to the increase gained at the same time by the live force, so that *the sum of the potential and the live force keeps an unchanged value through the whole duration of the motion.*

An isolated system has the conditions required for this theorem to be applicable; the only forces are then those that the various parts of the system exercise upon each other, and we have admitted that they would be derived from a potential; in this case one often gives to the potential for the internal forces the name of *potential energy* of the system; to the live force, the name of *live*[2] *energy*, the *real*[3] *energy*, or the *kinetic*[4] *energy*; to their sum the name of *total energy*; the previous proposition then takes the following form: *In the motion of a material system removed from the action of every external body, the total energy of the systems keeps an unchanged value*; under the name of the *Principle of the Conservation of Energy* this proposition has played a leading rôle in the development of Physics.

---

[1] Lagrange: *Mécanique analytique*, (2nd edn), Part II, Section III, §v.
[2] *Fr. vive.*
[3] *Fr. actuelle.*
[4] *Fr. cinétique.*

If the system studied were to be subjected to the action of certain external bodies, the value of its total energy could vary; the increase suffered by this energy during a certain lapse of time would be exactly equal to the work performed at the same time by the forces which arise from external bodies.

Using the equation for the active force in the case where the forces which operate are derived from a potential, Lagrange discovered an extremely important theorem regarding the *stability of equilibrium*.

Let us take a mechanical system subject to such forces, and without imparting to it any initial velocity let us place it in a state where the potential of the forces acting upon it is smaller than every neighbouring state; the laws of Statics show without difficulty that the system will remain in equilibrium in this state. The laws of Dynamics and, in particular, the live force equation, give us a new bit of information: At a given instant let us displace the system by a very small degree out of its state of equilibrium and let us impart to it some very small velocities; the system is going to be set in motion, but the various states through which it will pass in the course of this motion will always stay very close to the initial state of equilibrium, and the velocities of its different parts will keep to very small values; the initial equilibrium will be a stable equilibrium. Lagrange[1] gave a proof of this beautiful proposition which, by degrees of modification, Lejeune-Dirichlet[2] made quite rigorous.

In the neighbourhood of such a position of stable equilibrium, the system when slightly displaced from its state of equilibrium performs small oscillations; these oscillations result from the superposition of as many simple vibrations as there are independent variables[3]; the methods thought up by Lagrange for studying these oscillations are equally invaluable to the physicist and the engineer; they are no less important in Acoustics than in the theory of the vibration of machines.

Can a system be in stable equilibrium only in the positions where the potential reaches a minimum value? Lagrange believed to have proved this proposition; but his arguments were plainly insufficient; it is only in these present days that Lyapounov and Hadamard have been able, in a very extended case, to substitute some convincing deductions.

---

[1] Lagrange: *Mécanique analytique*, (2nd edn), Part I, Section III, §V, n° 25.

[2] Lejeune-Dirichlet: Über die Stabilität des Gleichgewichtes, *Crelle's Journal*, Vol. XXXII, §85, (1846).

[3] Lagrange: *Mécanique Analytique*, (2nd edn), Part II, Section VI, §I.

CHAPTER VIII

LAGRANGE'S ANALYTICAL MECHANICS

AND POISSON'S PHYSICAL MECHANICS

The notion of fictitious constraint force is that which distinguishes Lagrange's Mechanics from the Mechanics of Newton and Boscovich; in the latter, in fact, bodies are composed exclusively of *free* material points, so that all the forces that are considered actually are forces that are acting; in the former, contrary to this, bodies are continuous media whose various elements, mutually impenetrable, are mutually impeded in their motions.

Can one pass from the notion of constraint force introduced to Mechanics by Lagrange and recover all the results of this geometry by combining the bodies with the material points which are attracted to them? Laplace seems to have been the first to offer this opinion: "All terrestrial phenomena", he said[1], "related to molecular attractions, depend upon this kind of force, just as celestial phenomena depend upon universal gravitation. It now seems to me that their consideration has to be the principal object of Mathematical Philosophy. It seems to me even to be useful to introduce it into the proofs of Mechanics by abandoning abstract considerations of massless lines, whether flexible or not, and of perfectly hard bodies. Some attempts have made me see that in thus approaching nature one could give to these proofs as much simplicity and considerably more clarity as by the methods used up to today."

The numerous memoirs of Poisson were going to transform this remark into a veritable doctrine, a rival of that of Lagrange which was to strive to supplant it. The debate between these two methods was one of the most serious and, at the same time, one of the subtlest that the historians of mechanical explications have to relate.

In the first place let us remark that between the volume elements of a continuous medium, when treated by Lagrange's method,

---

[1] Lagrange: *Mécanique céleste*, livre XII, Chapitre I.

one can definitely admit the existence of these attractive or repulsive forces introduced by Newtonian Physics and called molecular actions. For example, when Gauss[1] treats a fluid as a continuous medium whose various elements are subject to such forces, and when he determines the shape of this fluid by the process of virtual displacements, he writes nothing that does not agree very exactly with the rules given in *Mécanique analytique*.

But the existence of these mutual actions does not prevent each part of such a continuous medium from being impenetrable from the neighbourhing parts, so that the presence of all these parts presents an obstacle to the motion of the continuous parts and constitutes a *constraint* for them.

This is the consideration of such constraints as are connected with the general notion of pressure within an arbitrary medium, solid or fluid, mobile or not; to define this notion it suffices to understand, after Lagrange, what has been called hydrostatic pressure.

Consider a continuous medium whose various volume elements are mutually impenetrable; the motion of each of the parcles of this medium is subject to certain constraint conditions which follow from the impenetrability of the adjoining parcels; in thought, isolate a portion of this medium, surrounding it by a closed surface; now remove it from all the remainder of the medium *whilst preserving the real force which acts on each of the elements of the portion isolated thus*; the constraints to which this portion is subject have changed through this operation, whilst the real actions which act upon it are kept unaltered; these forces, in general, no longer impart to it the motions that they imparted when they were situated in the middle of the medium; if one were to want the motion of this mass to be kept unaltered by the operation which isolated it, it would be necessary to adjoing to the forces actually acting, which were acting upon it, and that still act upon its elements, new forces which will be the *constraint forces*.

By Lagrange's methods, linked with the fundamental principles of the Calculus of Variations, he shows that these forces are applied exclusively to the surface that bounds the isolated mass; that each element of this surface supports a force of the same order of size as its area; that, in order to know the magnitude and direction of the force supported by an element, it is not necessary to know the surface of which this element forms a part, but only the position of the element within the medium; thus there is nicely defined the notion of pressure at each point of the medium and for each orientation of the surface element governed by this point.

When, for defining pressure within a body, one isolates a part of this body from everything surrounding it, great care is needed, as we have indicated, to suppress none of the real forces acting

---

[1] C.F. Gauss: Principia generalia theoriae figurae fluidorum in statu aequilibrii, *Commentationes Societatis Gottingensis recentiores*, Vol. VII, (1830); Gauss: *Werke*, Vol. V.

upon this particle. If, for example, one regards certain of these forces as originating from the neighbouring portions of the medium, and of such a kind that the suppression of these portions implies the disappearance of these forces, it will be assumed they will be replaced by other forces equal to them, originating from bodies not continuous to the isolated portion, and therefore producing no constraint upon its motion.

But it would be necessary to guard against simply saying, and without precautions, that the pressures are the forces it is necessary to apply to a portion of the medium, isolated from its surroundings, to give it the motion it would have in its natural place in the body of the medium. In fact under these conditions the pressures would replace not only the *constraints* arising from the presence of the parts of the medium contiguous to that we isolated, but even the *real forces* that the first parts are able to exercise upon the latter. This confusion does not seem to have been avoided by Lamé[1].

For Poisson, as for Boscovich, bodies were only continuous in appearance; in reality they were formed of isolated material points. If we were to consider a part of a medium, that is to say a group of material points, its infinitesimal virtual displacements would suffer no influence arising from the material points neighbouring this group without touching it; if we remove these neighbouring material points we suppress no constraint upon the group we have kept; but *we suppress the molecular actions that this group suffers arising from the material points we have removed*; the pressures that we are going to apply to the material points kept will have the object of compensating exactly the effect of these molecular forces that have been destroyed. According to this way of looking at things, the pressures are no longer constraint forces; they are the results of molecular actions exercised upon a part of the material points that compose the systems by the other material points of the system.

Such is the meaning attributed by Poisson to the *pressure* that is encountered in the study of solid or fluid media, the tension of a thread or a membrane.

It is, in fact, in his *Mémoire sur les surfaces élastiques*[2] that, for the first time, Poisson defined the tension of a membrane using this method; but soon he developed the consequences of this method in all parts of Physics, in the studies of Elasticity[3], Hydrostatics[4] and Capillarity[5]. According to him this in-

---

[1] Lamé: *Leçons sur la théorie mathématique de l'Élasticité des corps solides*, (2nd end), p. 10.

[2] Poisson: *Mémoire sur les surfaces élastiques*, read at the Institut, 1st August, 1814.

[3] Poisson: *Mémoire sur l'équilibre et le mouvement des corps élastiques*, read at the Académie, 14th April, 1828.

[4] Poisson: *Mémoire sur l'équilibre des fluides*, read at the Académie, 24th November, 1828.

[5] Poisson: *Nouvelle théorie de l'action capillaire*, (Paris), (1831).

novation constituted a principal reform, the creation of a new Mechanics, *Physical Mechanics*, called upon to supplant Lagrange's *Analytical Mechanics*. Here are the terms in which it is expressed at the end of the preamble of his *Mémoire sur les corps élastiques*:

"Let us add that it would be desired that geometers should reconsider the principal questions of Mechanics from this physical point of view conforming to Nature. It has been necessary to treat them in a way far too abstract for the discovery of the general laws of equilibrium and motion; and in the genre of abstraction Lagrange travelled as far as one can imagine when he replaced the physical bonds of the bodies by equations between the coordinates of their different points; the former constitutes *Analytical Mechanics*; but apart from this admirable conception, one could now raise up *Physical Mechanics* whose unique principle would be to remove all molecular interactions which transmit from one point to another the action of the given forces and that are the intermediary of their equilibrium. In this way one would have no more special hypotheses to make when one might wish to apply the general rules of Mechanics to some particular questions. Thus in the problem of the equilibrium of flexible bodies, the tension introduced to solve it will be the immediate results of the mutual action of the molecules displaced ever so little from their natural positions; in the case of an elastic strip, the moment of elasticity by bending will originate from these same actions considered in the whole thickness of the sheet, and its expression will be determined without any assumption; finally, the actions exercised by the fluids in their interior and on the walls of receptacles which contain them are also the resultants of the actions of their molecules on the surfaces pressed, or rather on an extremely thin fluid bed in contact with each surface."

Thus, according to Poisson, there exist two ways in which to conceive of Mechanics: in the one, which is that of geometers, the systems studied are subject solely to external forces, or to mutual attractions depending upon universal gravity, but they are subject to constraints; in the other, that of Physicists, the systems are formed from free material points; but to the real forces which the first Mechanics considered, it is necessary to add the molecular actions which are exercised between each pair of points; *these two Mechanics are equivalent for him who only takes account of their consequences; but the second clings more closely to the intimate nature of things.*

This doctrine of Poisson, as we have said, is only the development of a thought of Laplace; we should therefore not be astonished to rediscover it in the writings of the contemporaries of Poisson, particularly of those who founded the Theory of Elasticity. Pressure is defined according to Poisson's method at the beginning of the Mémoire where Navier[1] proposes, for the first time,

---

[1] Navier: *Mémoire sur les lois de l'équilibre et du mouvement des corps solides élastiques,* read at the Académie des Sciences,

the equilibrium conditions of an elastic solid. Cauchy[1] followed the same route when he extended the results obtained by Navier to non-isotropic bodies; in his numerous and important researches in elasticity he followed now Lagrange's method and now the method of Laplace and Poisson: "In the investigations of the equations that express the equilibrium conditions or the laws of the interior motion of solid bodies of fluids", he said[2], "one can consider these bodies as continuous masses whose density varies from one point to another by imperceptible steps, or as systems of distinct material points, separated from each other by very small distances". Cauchy seemed to apply himself to establishing the equivalence of the two methods in all circumstances.

Up to the present day the most distinguished minds have not ceased to propound, on the subject of pressures, the ideas expressed by Poisson, in admitting their equivalence with Lagrange's opinions, nay, in truth, to extol them as more in conformity with the veritable constitution of bodies than the latter.

In speaking about the theory of capillarity given by Poisson, J. Bertrand puts it thus[3]: "It is quite true that in physical and compressible fluid pressure cannot be distinguished from the resultant of molecular forces, and must be calculated, as Poisson remarked so often, by means of the function that represents them. But from the abstract point of view which geometers take, this pressure forms a force apart, different from the nature of those which he introduces so often in Mechanics under the name of *constraint forces*...".

De Saint-Venant, whose immense works have greatly contributed to the progress of elasticity, never ceased to defend the way in which Poisson viewed things. In the margin of a copy of *Mécanique analytique* which belonged to him, beside the passage where Lagrange showed so nicely that hydrostatic pressure is a constraint force, we find this note in his own hand: "Pressure is the averaged repulsion of fluid molecules". Some lines further down, in respect of a theorem, owed to Euler, on hydrostatic pressure: "This is still an *analytical* proposition, it would be desirable to convert it, as well as the others, into physical principles". Moreover, in the translation of Clebsch's *Traité de l'Élasticité*, de Saint-Venant devoted a long note[4] to the exposition and defence of Pois-

---

14th May, 1821.

[1] Cauchy: *Recherches sur l'équilibre et le mouvement intérieur des corps solides ou fluides, élastisues ou non-élastiques*, communicated to the Académie des Sciences, 30th September, 1822; *Bulletin de la Société philomatique, année 1823*, p. 9.

[2] Cauchy: Sur les équations qui expriment les conditions d'équilibre ou les lois du mouvement intérieur d'un corps solide élastique ou non-élastique, *Anciens Exercices, 3me année*, (1828), p. 160.

[3] J. Bertrand: Mémoire sur la théorie des phénomènes capillaires, *Journal de Liouville*, vol. XIII, (1848), p. 195.

[4] Clebsch: *Théorie de l'élasticite des corps solides*, (translated by Barré de Saint-Venant and Flamant), (Paris), (1881), p. 63 *et seq.*.

son's ideas.

J. Boussinesq[1], a faithful disciple of de Saint-Venant, never considered constraint forces in Mechanics, but only the resultants of molecular interactions.

In his remarkable *Traité de Mécanique rationelle*, de Freycinet[2] followed Poisson's idea very closely; he studied in parallel the systems he calls *geometric*, whose different points are united by constraints as understood by Lagrange, and the systems he called *dynamical*, whose points, free of any ties, exercise attractions or repulsions upon each other: "In Nature", he said, "*there are no geometric systems*".

We would never be finished if we wished to enumerate all the authors who, explicitly or implicitly, abandoned the notion of constraint force defined by *Analytical Mechanics* to adhere to the principles of *Physical Mechanics*.

These two methods appropriate to treating accurately the problems of Mechanics are both clearly and exactly formulated; that it is logically permissible to follow one or the other is something that could not be contested. On the other hand, what is legitimately contestable is the equivalence of the two methods; this equivalence, if it exists, would not pass as evident; it begs for a proof; it is necessary to prove, and not to postulate, that under all circumstances these two Mechanics lead to the same consequences. Therefore if by one of these methods one obtains results which do not agree with the other, one should not be scandalised at this contradiction; but by comparing experimentally the different results of the two methods, one could investigate which is the one best adapted to the facts.

The history of the theory of capillarity provides us with an occasion for applying these remarks.

Ever since Newton the majority of geometers have been agreed upon attributing the shape taken by a fluid in a narrow container to the molecular attractions which the various parts of the fluid exercise upon each other. This hypothesis agrees, it goes without saying, with the principles of Poisson's Mechanics; but it is equally reconcilable, as we have remarked, with the principles of Lagrange's Mechanics; only in this latter is the fluid supposed continuous; the molecular attractions then are exercised not between points but between infinitesimal volumes in addition to these interactions, one will then have to consider the constraints of contiguous elements; moreover, to these constraints one will very logically have to add others, such as the *incompressibility* condition, imposing an unchanging volume upon each elementary mass.

Lagrange's processes allow the study of the equilibrium of such fluids. One can, as for example Gauss[3] did, impose a vir-

---

[1] J. Boussinesq: *Leçons synt étique de Mécanique générale*, (Paris), (1889).

[2] de Freycinet: *Traité de Mécanique rationelle*, vol. I, (Paris), (1858), p. 240.

[3] C.F. Gauss: Principia generalia theoriae fluidorum in statu

tual modification upon the whole system, which avoids considering
the pressure in the body of the fluid; one can also, as with
Franz Neumann[1], introduce this pressure into the calculations by
following very closely the method employed by Lagrange in hydrostatics; the results obtained by one or the other of these two
processes agree completely with those Laplace had found[2]; the illustrious author of *Mécanique céleste* additionally made use of
the *Principle of the Equilibrium of Channels* thought of by Clairaut[3] and reduced by Lagrange to the Principle of Virtual Velocities.

In his turn, Poisson[4] approached the problem of the equilibrium
of liquids in capillary spaces according to the rules of Physical
Mechanics; the consequences he reached could not agree with the
propositions of Laplace and Gauss if the fluid were assumed to be
incompressible; to recover the laws of capillary phenomena such
as had been announced by the author of *Mécanique céleste*, it is
necessary to assume that the liquid is compressible and that its
density varies very rapidly in the neighbourhood of end surfaces.

This disagreement Poisson turned into an objection against the
theory of Laplace and Gauss; by disallowing compressibility in a
liquid, they had "omitted a physical circumstance the consideration of which was essential, and without which capillary phenomena would not take place".

This conclusion which Poisson drew from his investigations is
incorrect; the only legitimate conclusion that could be drawn
would be formulated in these terms: Incompressible fluid, logically conceivable in Analytical Mechanics, is inconceivable in
Physical Mechanics. "In fact", remarked Quet[5], "no account is
taken of the constraint forces which one is still obliged to admit if one wants liquids, assumed incompressible, to be capable
of supporting the more or the less strongly their elements against
one another, and of transmitting pressures to the interior. The
suppression of these constraint forces makes not only capillary
phenomena disappear, but also all of Hydrostatics and Hydrodynamics, and there is no need of calculations to see it. Without
them the equilibrium conditions necessarily are incomplete, and
there would be reason to be suprised that one was not led to flag-
grant contradictions by a method that takes no account of all the

---

aequilibrii, *Commentationes Societatis Gottingensis recentiores*,
vol. VII, (1830); Gauss: *Werke*, vol. V.

[1] F.E. Neumann: *Vorlesungen über die Theorie der Kapillarität*,
(Leipzig), (1894), Kapitel VIII.

[2] Laplace: *Supplément au Xe livre de la Mécanique céleste; sur
l'action capillaire.—Supplément à la théorie de l'action capillaire*.

[3] Clairaut: *Théorie de la figure de la Terre*, (Paris), (1743).

[4] Poisson: *Mémoire sur l'équilibre des fluides*, read to the Académie des Science, 24th November, 1828; *Nouvelle théorie de l'action capillaire*, (Paris), (1831).

[5] Quet: *Rapport sur les progrès de la Capillarité*, (Paris),
(1867).

causes".

Analytical Mechanics and Physical Mechanics are therefore far from leading to equivalent results under all circumstances. Since they are different, which is the most suitable one to adopt? Is Physical Mechanics, as it pretends to be, the one that by the most natural and shortest paths is the most exactly modelled on facts?

First of all, let us remark that in order to take its calculations as far as possible, sooner or later it will be necessary to treat bodies as assemblages of free material points and to restore to matter the continuity which it had been denied. Under this condition alone one can transform into integrals, easy to handle, the sums, in analysis, with which its processes provide it in the first place. This transformation of sums into integrals is obtained not without discussions that are all onerous, nor without approximations that are often coarse; in this operation mathematical rigour suffers almost as much as elegance; each seems to recommend the calculations of Analytical Mechanics. But other difficulties ruffled the path of Physical Mechanics.

Let us consider an assemblage of free material points; let us assume that between any two of these points there exists a reciprocal action proportional to the product of the masses of these two points and a function of the distance separating them. First of all let us imagine that however small this distance may be, this action is attractive. It is clear that the system, when shielded from every external force, could not be in equilibrium; all the inner material points would tend to be united into a single one; it would be the same *a fortiori* if a uniform pressure were to be exerted upon the surface of the body; the latter would then have to have zero volume and infinite density.

Boscovich clearly had perceived this difficulty. To ward it off he assumed that the reciprocal action between two points would always become repulsive when their mutual separation fell below a certain limit. From this same remark Navier and Lamé had been led to modify more profoundly the very principles of Newtonian philosophy; according to these physicists, when the body is in the *natural state*, that is to say shielded from every external action and being in equilibrium, two arbitrary points exercise no action upon each other; their reciprocal action only originates from the effect of a deformation that increases or decreases the separation of these two points; it is proportional to the alteration occurring in the distance between these two material points and always tends to oppose this alteration; furthermore, its magnitude depends upon the original distance between these two particles. This view has met few supporters; moreover, it does not avoid certain grave objections which Poisson's theory runs into, and upon which it remains for us to say a few words.

Let us first observe this: When the existence of constraints is denied, when bodies are regarded as assemblies of free material points exercising attractive or repulsive force upon each other, it is impossible to introduce in a logical way a line of demarcation between, on the one hand, isotropic elastic solids, and compressible liquids on the other; everything that will be proved for isotropic elastic bodies will have to remain true, in particular, for compressible liquids.

Now, the study of isotropic solids led Poisson to some remarkably simple consequences; thus, when a prism formed of such material is stretched, the ratio of the transverse contraction to the longitudinal extension is fixed and equal to one quarter; or yet again, in every isotropic body, the ratio of the bulk modulus to the modulus of elasticity is equal to two thirds.

Does experiment verify these conclusions? With an extravagance of effort, Kirchoff found them to be exact in certain particular cases; but according to Wertheim they do not hold for metals. Consequently, "a solid body, even an isotropic one[1], cannot be considered as formed from a system of molecules that are mutually attracted or repelled as a function of distance ... without being subject to certain constraints such as are considered in Analytical Mechanics".

The supporters of Poisson's theory, it is true, would always be able to oppose a conclusion of not accepting experimental contradictions by declaring that bodies whose properties did not accord with their formulae are not truly isotropic, but are built up from an entanglement of crystals; they have not been unsuccessful in using this loophole; but they can be opposed by an argument to which there seems to be no reply.

Everything Poisson's theory stated for isotropic elastic bodies must, in all logic, be equally applicable to liquids. If, therefore, for truly isotropic bodies the bulk modulus is obtained by multiplying the modulus of elasticity by three halves, this proposition must remain true for liquids. Now, this cannot be, for the bulk modulus differs from zero, whilst the modulus of elasticity is zero.

Hence it is impossible to retain the principles upon which Poisson wanted to base Physical Mechanics, unless recourse is had to subtleties and subterfuges. Moreover, Poisson had already been seen to be reduced to these desperate means of defence; to be convinced of this it suffices to read the *Notions préliminaires* by which is opened the *Mémoire sur l'équilibre des fluides*. There Poisson no longer regards the elements of bodies as points without extension, not only does he treat them as shaped particles, but he invokes, furthermore, under the name of *secondary action*, a force that depends upon the shape of the molecules which hampers or eases their mobility, and to which he attributes all the effects that Analytical Mechanics would have attributed to constraint forces.

When a theory thus multiplies ruses and chicanery to defend itself, it is useless to pursue it, for it has become elusive; but it would then be idle to grapple with it, for to every right-

---

[1] É. Mathieu: *Théorie de l'élasticité des corps solides*, vol. I, (Paris), (1890), pp. 6,39.

thinking mind it would be a vanquished doctrine. Such is Physical Mechanics.

Over the difficulty upon which the latter had just broken itself, its rival, Analytical Mechanics, triumphed effortlessly; its methods, put to work by Cauchy, Green and Lamé, show that the elastic properties of an isotropic body depend upon two distinct coefficients quite independent of each other; these coefficients Lamé denoted by the two letters λ and μ. In a tensioned prism the ratio of the transverse contraction to the longitudinal dilatation has the value $\frac{1}{2}\lambda/(\lambda + \mu)$; the ratio of the bulk modulus to the modulus of elasticity has the value $(\lambda + \mu)/3\mu$; for various substances these two ratios can therefore take the most diverse values; the values allowed by Poisson can be recovered if it is assumed that the two coefficients λ, μ are equal to each other; but this assumption cannot be made in general, since for liquids λ is zero whilst μ has some positive value.

CHAPTER IX

THE KINETIC THEORY OF GASES

Analytical Mechanics, triumphant, is not constructed entirely from *shape* and *motion*, the sole elements to have been admitted by Cartesians in the explication of the world; like the atomists, it was not satisfied with adding *mass* to these elements; in addition it invoked the idea of *force*; but these four notions sufficed it for constructing a system admirable in its fullness and logical unity. This system realised Leibniz's dream; it was therefore, as that great metaphysician had recognised, a reaction contrary to the trends of Gassendi, Descartes and Huyghens, a return to the doctrines of the Old School.

The continual ebb and flow that made human opinion oscillate had pushed the Mechanics of Lagrange and his contemporaries towards the ancient Peripatetic Physics; with the ebb-tide succeeding the flow, the science of Nature was now going to drift towards atomistic doctrines.

This change of direction in the current that bore the theories of Physics had been determined by the discovery of the equivalence between heat and mechanical work. This discovery, as we shall see in the Chapter following, agreed very well with the hypothesis that heat is a motion, a hypothesis postulated by Descartes and accepted by all physicists who preceded Black and Crawford; it was therefore naturally called upon to bring back into favour Cartesian or atomistic Physics, the explications of which rejected the notion of force.

Amongst these explications the atomistic theory of the properties of gas first of all attracted the attention of physicists. This preference was, so to say, forced, for the laws relating to gascous bodies were precisely those that had provoked the creation of Thermodynamics, those which lent themselves to the easiest and most complete calculations.

Having been prepared by the essays of Leibniz, Malebranche, James Bernoulli, Parent, and John Bernoulli I, the doctrine known today under the name of the *Kinetic Theory of Gases* was defined precisely in 1738 by Daniel Bernoulli in the Tenth Section of his *Hydrodynamics*[1].

Let us imagine, he said, a cylindrical container with vertical generatrices, the upper orifice of which is closed by a piston loaded by a certain weight. Let us fill this vessel with a multitude of very small corpuscles, agitated in all directions; these corpuscles, striking the piston with repeated blows, prevent it from descending; if one increases the weight loading the piston, the latter will descend until the small bodies, confined in a smaller space, can sustain it by their now more frequent impacts. We have before our eyes a mechanism that simulates the most obvious characters of an elastic fluid; would there be any point in explaining the properties more exactly?

Let us assume that the gaseous particles are perfectly elastic spheres, *all moving with the same speed*; in addition, let us imagin that they are so small that the volume actually occupied by these particles is negligible with respect to the volume within which they move, at least when the air is in the usual atmospheric conditions; finally, let us admit that in two circumstances where this air is equally hot, these particles move equally fast. We find without difficult that in different masses of air equally hot the pressure is proportional to the density, in agreement with the observations of Boyle, Townley and Mariotte; however, this law would doubtless cease to be exact for very condensed air, for the volume occupied by the moelcules would become comparable to the apparent volume of the gaseous mass[2].

If one carries a mass of gas of a specified degree of heat to another degree, also specified, the speed of the molecular motion changes from one value to another; at constant density the increase of the pressure is proportional to the increase in the square of the velocity; thus one recovers this proposition[3] that Amontons had obtained experimentally in 1702: *In different masses of air of differing densities, but equally hot, their elasticities are as the densities; the increase in elasticity owed to a specified increase of heat is proportional to the densities.*

"Knowing[4] some values proportional to the elasticities displayed in different circumstances by the same mass of air enclosed in the same space, it is easy for us to measure the degree of heat of this air, provided that we were to adopt a conventional definition of a double, triple, *etc.*, degree of heat; a definition which is arbitrary and in no way imposed by the nature of things; one can, so it seems to me, take for the measure of the degree of heat the elasticity of a mass of air whose density is always equal to the usual density".

---

[1] Danielis Bernoulli: *Hydrodynamica, sive de viribus et motibus fluidorum commentarii*, (Argentorati), (1738).
[2] D. Bernoulli: *loc. cit.*, p. 202.
[3] D. Bernoulli: *loc. cit.*, p. 203.
[4] D. Bernoulli: *loc. cit.*, p. 204.

The *temperature scale* adopted here by Daniel Bernoulli is that which Amontons had proposed in 1702 and for which he had constructed a thermometer; it coincides with that which today provides *absolute temperatures*. On the condition that this scale is used, under all circumstances the air exerts a pressure proportional to the product of its density with the absolute temperature.

The powerful attempt by which Daniel Bernoulli tried to render an account, in accordance with the principles of the atomists, of the laws of compressibility and dilatation of gases was completely forgotten when Krönig[1] and Clausius[2] recovered the essential ideas, and which the latter interpreted in three fundamental Mémoire[3], and drew from them a detailed explication of the phenomena offered by gases.

The assumptions of Clausius in his first Mémoire are almost identical to those Daniel Bernoulli had formulated. Gases were to be formed of hard spheres whose diameters were very small compared to the mean value of the distance separating two neighbouring spheres; each sphere moved in a straight line with uniform motion until it met a wall or another sphere; it then rebounded in accordance with the laws of impact of elastic bodies; these laws imply variations of speed for the bodies struck; the elastic spheres which constitute the gas therefore cannot all move with the same speed, as Daniel Bernoulli had desired, whose analysis must be modified at this one point; it is no longer the uniform speed of molecular motion which is independent of all conditions save temperature; this character now belongs to the average kinetic energy; it is the latter that one can take as measure of the absolute temperature.

But in Clausius's second Mémoire, the hypotheses of the Kinetic Theory of Gases lose this simplicity which reconciled them with the principles of Atomistic Physics; between two gaseous molecules a reciprocal action is assumed which accords very exactly with the rules proposed by Boscovich: attractive when the mutual distance of the two molecules is not of the same order of magnitude as their own dimensions, it becomes energetically repulsive when this distance falls below a certain limit; later Maxwell was to make precise this latter assumption by admitting that the repulsive action is inversely proportional to the fifth power of the distance.

From that not only the foundations of the Kinetic Theory of Gases become more complex, but they change character. Physics, the Atomistic Physics that had been believed to be triumphant, was forsaken anew. The existence of molecular forces was admit-

---

[1] Krönig: Grundzüge einer Theorie der Gase, *Poggendorff's Annalen*, vol. XCIX, pp. 315,1856.
[2] Clausius: Ueber die Art der Bewegung, welche wir Wärme nennen, *Poggendorff's Annalen*, vol. C, p. 353, 1857.
[3] These three mémoires, published in 1857 to 1862 in *Poggendorfs Annalen*, have been translated into French by F. Folie in: R. Clausius: *Théorie mécanique de la Chaleur*, vol. II, (Paris), (1869).

ted by Clausius and Maxwell, just as it had been by Boscovich and Poisson.

But in simple comparison with Poisson's Physics, the new doctrine offers great complications.

For Poisson's School, a gas whose density and temperature appear unchanged to our senses and instruments is actually a gas in equilibrium; at each of the point masses which compose it, all the forces are exactly counterbalanced, and this point remains at rest. For the kinetic theory, the equilibrium we observe is only an apparent equilibrium. If it were given to us to perceive molecules or atoms, at the site of this apparent rest we would behold a tumultuous agitation, a chaos of wild courses and incessant impacts. A space which seemed imperceptible to our eyes, even armed with a more powerful microscope, would appear in our new viewpoint as an immense extent; a duration of a very small fraction of a second would seem as long as an hour to a mind capable of following the progress of the atoms. If in such a space and during such a time we counted the atoms that went in a certain direction with a certain speed, and those which went in the contrary direction with the same speed, we would find that the very large number of the first and the very large number of the second differed from each other by a number that is not very great. It is this approach to equality, it is this balance between the chances that the molecules have of being thrown in a particular direction and the changes they have of being thrown back in the opposite direction, which constitutes the apparent state of equilibrium of the gas. Thus the population of a country is stationary when each year the number of births hardly differs from the number of deaths, and that from one year to another the gap between these two numbers changes sign. According to Maxwell's felicitous expression, the equilibrium of a gaseous mass is a *statistical equilibrium*

These simple pointers announce sufficiently the extreme difficulties that were going to be met by physicists when they wanted to use the kinetic hypotheses as the point of departure for rigorous deductions; these difficulties are summarised in these two words: *approximation, probability*.

These hypotheses put disordered motion and discontinuous multitude into the uniformity and continuity that our sense perceive and our instruments measure. These are sums of an immense number of terms following in an irregular manner, which will be provided for the mathematician; to recover the quantities that are accessible and which are only mean values, he will have to transform these sums into integrals; in the course of these transformations he will have to take minute account of the order of magnitude of these elements, at one and the same time very small and very numerous, that will have to be considered incessantly; it will be necessary to appreciate exactly which terms are small enough to be neglected and which ones are large enough to be kept; it will be necessary to determine the degree of approximation with which each sum is represented by the integral that has been substituted for it.

Poisson's Physical Mechanics and already recognised these difficulties; for the geometer who discussed these kinetic hypotheses, they are not the most formidable ones.

What our senses take as a true state of equilibrium is a state of statistical equilibrium only, a state which remains stationary on the average, because the chance blows that tend to disturb it in one direction are compensated by the chance blows which tend to disturb it in another. So when we wish to know if a certain distribution of atoms and motions represents an apparent state of equilibrium, a state capable of lasting, we shall have to reckon the changes that are in favour of each of the causes capable of disturbing it. From there we are now obliged to have recourse to the *Probability Calculus*, despite the hesitations and doubts that seem inherent in this order of argument.

The least problem of kinetic theory will therefore be a difficult riddle to unravel, even difficult to state, if one seeks to satisfy the requirements of rigorous minds; the most zealous supporters of this doctrine willingly acknowledge that it is hard to discuss it in an irreproachable way. "The problems thus posed for mathematicians", said Brillouin[1], "are of a formidable complexity; but is it not evident that this complexity lies in the nature of things, and that a very simple fundamental idea might only serve to group together a very large number of phenomena if the logical analysis of the content of this simple idea were to lead to a great richness of associations and combinations? Now this richness is possessed by the molecular hypothesis; the rigorous translation of it into mathematical language is extraordinarily difficult; instead of assuring each step, it is necessary at each moment to jump over an abyss; it is not upon a sound national highway that we advance, it is upon a glacier bristling with cracks and criss-crossed with crevasses. For it has to be acknowledged that the elegant arguments do not all seem to be very sound; and certain quite sound arguments are discouragingly long".

Daniel Bernoulli believed that all molecules of which a gaseous mass is composed move with the same speed; this assumption is plainly inadmissible; from one molecule to another the velocity differs in direction and in magnitude. How are these various velocities distributed amongst the molecules in the heart of a mass in apparent equilibrium? Evidently this is the first question that the Kinetic Theory of Gases has for examination. It can be stated with more precision in the following way: Perfectly elastic molecules are shot in very large numbers into a space that is very large in comparison with the volume they actually occupy; between these molecules there are exercised actions that are attractive or repulsive in conformity with the principles of Newtonian Philosophy; the average kinetic energy, or in other

---

[1] M. Brillouin: Preface to *Leçons sur la Théorie des gaz*, by L. Boltzmann, (translated into French by A. Gallotti), (Paris), (1902), p. 14.

terms the temperature, is specified; in each direction in the space and at each instant, how many molecules are there moving with a velocity lying between two given limits?

Maxwell was the first to obtain a solution of this problem; the elegant rule he stated recalled that by which the method of least squares distributes the accidental errors made during a large number of observations in the determination of a quantity. But the first intuitions of Maxwell were not proofs; it needed great efforts to support them with rigorous arguments; in these efforts that great Scots physicist received a powerful aid from L. Boltzmann[1].

In order to prove Maxwell's theorem it suffices to formulate some extremely general hypotheses; but if one is restricted to these hypotheses, the consequences of the Kinetic Theory of Gases are too indecisive and too little defined for it to be possible to compare them with experiment. If one wishes to construct a theory capable of being subjected to the control of facts it is necessary to make the hypotheses more precise, to delimit and detail them with new assumptions, and these assumptions might vary according to the liking of the authors. From this there came various particular theories, mutually disparate, although all derived from the same general idea; they were discordant in their consequences, and offered only a partial agreement with the facts; from this there arose in this part of Physics a somewhat chaotic state which Brillouin[2] describes for us in these terms:

"The obligation to end up with some averaged results, the only observables, imposes the use of statistics and probabilities; but our present ignorance of the physical properties of molecules and of the law of molecular interaction gives rise to plenty of doubts about the correctness of the assumptions made in the course of calculations about the relative independence of various probabilities. Often it also seems impossible to pursue theory without adopting a particular law of interaction, say of impact, or of the $1/r^5$ repulsion; and yet in the final equations there certainly are some general characteristics that are independent of this law of interaction. Hence the difficulties are numerous; each author surmounts them as best he can. Uniform in its general ideas, the Kinetic Theory of Gases is diverse in its formulae; these are the very same general ideas that all authors have endeavoured to express in mathematical language; but the choice of simplifications, conscious or unconscious, which is required for putting the physical problem into the form of an equation has been justified by each author according to the old adage: *To translate is to betray*. Hence there are varying mathematical theories, and

---

[1] Under the title *Vorlesungen über Gastheorie*, (Leipzig), (1896-98), L. Boltzmann published an invaluable exposition of the Kinetic Theory of Gases. The first volume of this work has been translated into French by Gallotti, with a preface by Brillouin, (Paris), (1902).

[2] M. Brillouin: *loc. cit.*, p. 18.

it is a very delicate question to know whether on such and such a point the theory of the author so and so is merely imperfect or actually false".

It seems, however, that the most convinced supporters of the kinetic hypothesis, and in particular the illustrious Boltzmann, have renounced the reduction of this chaos to order and unity, and of drawing from this hypothesis, helped by a certain number of secondary assumptions, a coherent doctrine conforming to all the facts revealed by the study of perfect gases. They appear to be resigned to seeing in the various forms of kinetic theory only some mechanical examples[1], which *imitate* certain properties of gases, which may, by way of *analogy*, give useful indications[2] to experimentalists, but which *explain* in no way the real constitution of gases, which in no way prove that matter is actually formed as the atomists would have it. "In presenting the theory of gases as a set of *mechanical analogies*", said Boltzmann[3], "we are already indicating by the choice of this expression how distant we are from admitting, in a final manner and as a reality, that bodies are composed in all their parts of very small particles."

---

[1] L. Boltzmann: Lecons sur la Théorie des gaz, vol. I, (translated by A. Gallotti), (Paris), (1902), p. 151.
[2] L. Boltzmann: *loc. cit.*, p. 171.
[3] L. Boltzmann: *loc. cit* , p. 4.

CHAPTER X

THE MECHANICAL THEORY OF HEAT

Amongst all the substances of which Physics studies the compression, dilatation, heating and cooling, the group of perfect gases is distinguished by the uniformity and simplicity of its properties. Now, when it was proposed to explain these properties by invoking only the shape of the atoms, their motions, and their mutual interactions, one ran into some obstacles that were difficult to get over; despite the efforts geometers and physics had lavished upon it, the Kinetic Theory of Gases was seen to be very nearly constrained to renounce its first pretensions; it no longer dared to be given out as explaining the nature of gaseous substances; it was content to imitate them, to represent them.

If the Kinetic Theory of Gases had seen its development stopped by insurmountable barriers, if it had had to deviate from the direction that it was assigned first of all, then with all the more reason would we have met the same obstacles and ascertained the same deviation by studying the much vaster doctrine that pretended to explain by shape, motion and force all the accompanying phenomena of a release or absorption of heat; this doctrine was that which had come to be called the Mechanical Theory of Heat.

It is necessary to go back to Descartes to recover the origin of the hypothesis that put the cause of our sensations of hot and cold in a lively and disordered agitation of small parts of bodies. Before him the Scholastics regarded hot and cold as qualities; the ancient atomists, and Gassendi himself, admitted the existence of special atoms that produced the sensation of heat, whilst other atoms generated cold. After Descartes, on the contrary, all physicists who were disciples of Huyghens or who quoted Newton as their authority, admitted that heat was an effect of molecular motion. This hypothesis reigned uncontested up to the last years of the eighteenth century; then only the calorimetric investigations of Black and Crawford rendered a momentary favour to some assumptions analogous to those Gassendi extolled; they made heat to be treated as a fluid to which the new chemical nomenclature was to give the name of *caloric*.

In 1783 Lavoisier and Laplace still hesitated between the new hypothesis that regarded heat as a fluid and the old Cartesian hypothesis which they stated[1], moreover, with great force and great precision: "Other physicists think that heat is only the result of indetectable movements of the molecules of matter. To develop this hypothesis we shall observe that in all the motions where there is no sudden change there exists a general law which geometers have called by the name of the *Principle of the Conservation of Kinetic Energy*[2]; this law consists in this, that in a system of bodies which act upon each other in an arbitrary way, the kinetic energy, that is to say the sum of products of each mass by the square of its velocity, is constant. If the bodies are set in motion by accelerating forces, the kinetic energy is equal to that of the origin of the motion, rather to the sum of masses multiplied by the squares of their speeds owing to the actions of the accelerating forces. In the hypothesis we shall examine, heat is the kinetic energy which results from the indetectable motions of the molecules of a body, it is the sum of the products of the mass of each molecule with the square of its speed.

"We shall not decide between the two preceding hypotheses; several phenomena appear to be favourable to the latter; such is, for example, that of the heat which the rubbing together of two bodies produces ... ".

In spite of the admirable researches of Laplace and Poisson, the triumph of the caloric hypothesis was short-lived; certain facts too clearly contradicted this doctrine; such was the release of heat in the friction of bodies, a release that was known from time immemorial and which Rumford had made particularly plain in the celebrated experiment at München; again, such was the observation of Gay-Lussac that a gas, upon expanding into a vacuum, neither absorbs nor releases heat. Moreover, the Optics of Young and Fresnel, in giving to light the character of a vibratory motion that had been attributed to it by Huyghens and Malebranche, put the doctrines of Descartes and his successors back into favour; it ruined the emissionistic hypotheses borrowed from the ancient atomists and Gassendi. Also Sadi Carnot had already written: "Heat is the result of a motion"; then, defining precisely the mechanical equivalent of heat, indicating the various methods which can serve for measuring it, he gave a first numerical evaluation of it.

Sadi Carnot died in 1832, but his notes remained unpublished until 1878, leaving to Robert Mayer the glory of first publishing in 1842 a definition and evaluation of the mechanical equivalent

---

[1] Lavoisier et Laplace: *Mémoire sur la Chaleur*, read to the Académie des Sciences, 18th June, 1783.
[2] *Fr.*: "forces vives".

of heat.

The discovery of Mayer was not inspired by the opinion that heat is a molecular motion, for the illustrious doctor from Heilbronn had rejected this assumption; on the other hand, this hypothesis was the stimulant of the researched pursued by his continuers Joule and Golding; it saturated the pages which in 1850 Clausius dedicated to the precise statement of the *Principle of the Equivalence of Heat and Work*.

Today this statement can be freed from every hypothesis touching the nature of heat. Let us recall this statement, and to avoid every useless complication let us agree to evaluate heat in *mechanical units*, that is to say, to multiply every quantity of heat by the mechanical equivalent of heat.

Each such state of the material system that it is proposed to study corresponds to a well defined value of a certain quantity, the *internal energy* of this system; when this system changes its shape or density, when it is heated or cooled, when it passes through one of the states solid, liquid, gaseous, to another, when it is the seat of a chemical reaction, when it is electrified or magnetised, its internal energy changes value; on the other hand, it remains the same when the system is at rest or in motion, when the speed of each of the parts which comprise it is small or large.

When the system suffers a modification, the kinetic energy and the internal energy each increase by a certain quantity; a certain release or absorption of heat is produced; finally, the forces that the outside bodies exercise upon the system carry out a certain work. *If we subtract from the external energy the increase in the kinetic energy and the increase in the internal energy, we obtain the quantity of heat released*. Such is the statement of the Principle of the Equivalence of Heat and Work.

Whatever may be the origin one may wish to attribute to this principle, whether one regards it or not as linked to the hypothesis that makes heat a form of motion, one must hold it as one of the most solid pillars of real Physics. If one wishes to reduce all physical phenomena to figure, motion, mass and force, one must first of all give a mechanical explication of the Principle of the Equivalence of Heat and Work.

Furthermore, the task is easy: the mechanical interpretation fo this principle is contemporaneous with its discovery: Helmholtz in 1847, Clausius in 1850, each formulated it quite precisely.

First of all let us consider a system that appears to be in equilibrium. The molecules which comprise it are provided with a motion that is of so small an amplitude that it is indiscernible; but this motion is one of very great rapidity; with every agitation of the molecules in all directions, in a disordered way, this motion leaves unchanged the mean state of the system, which is a state of *statistical equilibrium*. To these *stationary motions*, as Clausius called them, there corresponds a certain mean kinetic energy.

MECHANICAL THEORY OF HEAT

If the system under study no longer seems to be in equilibrium, but in motion, the molecules are no longer provided exclusively with stationary motions; the real motion which involves each of them is obtained by combining the stationary motion and the perceptible motion.

This real motion corresponds to a certain kinetic energy. In general, when one combines two motions it is not true that the kinetic energy of the resultant motion will be equal to the sum of the kinetic energies of the component motions; therefore it is not correct that the total kinetic energy of the system is, at every instant, the sum of the kinetic energies of the stationary motions and the kinetic energy of the perceptible motions.

But in the perceptible motions that we have for study, the speed of each material point varies gradually, in general; in a time that seems very short to our means of perception the variation of this speed is also very small; on the contrary, in this same time the speed, which in the stationary motion animates the same material point, has changed direction an immense number of times; when calculated for such an interval of time, the mean value of each of its components differs extremely little from zero; from there a very elementary proof allows us to affirm that the mean kinetic energy of the system, taken over the same time, is the sum of the kinetic energy of the perceptible motions and the mean kinetic energy of the stationary motions.

The molecules composing the system exercise attractive or repulsive actions upon each other; these internal actions admit a potential; thanks to the stationary motions which agitate the molecules, the value of this potential varies without ceasing, even in a system that appears to be in equilibrium; but in such a system it oscillates between very narrow limits about a mean value which characterises the state of statistical equilibrium of the system. If this system suffers a perceptible change, the interior forces perform a work that differs from the decrease suffered by this mean potential.

The external bodies which surround the system exercise certain actions upon it, and during a given lapse of time these actions perform a certain work.

First of all, this work comprises the work that it would be necessary to do to give, in the same time, the same perceptible displacement to the perceptible masses if they were not internally agitated by stationary motions; but it also comprises another thing; without analysing the nature of this second share, we can call it the *quantity of heat* that the system has *received* from the external bodies; by changing the sign of this quantity we shall have the *quantity of heat given up* by the system.

Dynamics provides us with the following theorem: The sum of the external and internal works is equal to the increase in the total kinetic energy of the system. Let us, then, use it, and we shall obtain the following proposition:

The sum of the external work and the decrease suffered by the perceptible kinetic energy is equivalent to the sum of three terms:

(1) *The quantity of heat given up;*

(2) *The increase in the mean potential of the internal actions;*

(3) *The increase in the mean kinetic energy of the stationary motions.*

It now suffices us to give the name *internal energy* of the system to the sum of the mean potential of the internal actions and the mean kinetic energy in order to recognise the statement of the Principle of the Equivalence of Heat and Work.

This principle is not the only one invoked in the theory of heat; the latter only attains its complete development in invoking another principle: the *Sadi Carnot and Clausius Principle*.

The assumptions about the mechanical nature of heat contributed nothing to the discovery of this latter principle; some postulates, that it had been possible to infer from the body of experimental truth, had led Sadi Carnot to state it in a form that implied the Caloric Hypothesis; later on Clausius modified it in such a way that it could agree with the Principle of the Equivalence of Heat and Work; the various statements which this great physicist gave it were independent of everything that had been attempted for explaining the properties of heat by the laws of force and motion.

These statements made temperature to play an essential rôle that gave this physical property quite a different face. In fact they postulate the existence of a certain quantity whose value is fixed for a particular degree of heat, in whatever body this degree of heat may be realised; this value is increased, moreover, in proportion as this body, whatever it may be, becomes hotter. This quantity is called the *absolute temperature*.

When a system suffers an infinitesimal modification it releases a certain quantity of heat which is, itself, infinitesimal, the quotient of this quantity of heat by the absolute temperature of this system is the *transformation value* of the infinitesimal alteration of state. A finite modification is a succession of infinitesimal modifications, each of which has a transformation value; the sum of these transformation values is the transformation value of the total modification.

These definitions permit the formulation of the Sadi Carnot and Clausius Principle; here is the most general statement of it:

The transformation value of a modification is equal to the decrease suffered, under this modification, by a certain quantity, related to all the properties which fix the state of the system, but independent of its motion. Clausius gave the name of *Entropy* of the system to this quantity.

The application of this principle to the perfect gases leads initially to a conclusion worthy of mention: The absolute temperature considered here is identical with the temperature that, in 1702, Amontons read on his thermometer; to that of which, in 1738, Daniel Bernoulli proposed the use; finally, to that which, in 1812, Desormes and Clément named *absolute temperature*.

MECHANICAL THEORY OF HEAT

Like Carnot's formula, the formula for the equivalence of heat and work can, as we have seen, be rendered safe from every hypothesis about the structure of matter and the nature of heat. Upon these two formulae, which leave the nature of heat undetermined, one can construct a complete body of doctrine, independent of the various systems of mechanical explications; this doctrine does not have any ambition to reduce all the phenomena it analyses to shape, mass and force; but in putting restrictions to its pretensions, it assured its deductions considerable security. Such is *Thermodynamics*, set up as an autonomous doctrine by Clausius and Kirchoff, and added to by continual discoveries.

Amongst physicists there are those who are content with knowing less, in order to know it better, who are resigned to ignoring the basis of things provided that phenomena may be described with precision and related to each other with rigour; the former accepted this restricted description of the theory of heat. But those who like to explain by mechanical arguments could not accept this form given to Thermodynamics as being definitive; for them it was only a means of proceeding towards the reduction of the laws of heat to the laws of motion.

Now, as we have seen, the Principle of the Equivalence of Heat and Work reduces without difficulty to the law of kinetic energy; in order to make Thermodynamics completely a chapter of Mechanics it suffices to draw from Carnot's Principle some theorems in Dynamics and some assumptions that have been made about the nature of heat; to start from these premisses it suffices that by dividing by the absolute temperature the quantity of heat released in an infinitesimal modification, one obtains the decrease of an Entropy as a function of the state, only, of the system.

Is the meaning, even, of the proposition to be proved, very exactly fixed? The interpretation of the Principle of the Equivalence of Heat and Work has sharpened the meaning that the Theory of Mechanics attributes to the quantity of heat given off by a system; but what combination of masses and motions must one substitute for absolute temperature?

When it is a question of perfect gases the Kinetic Theory leads to the identification of absolute temperature with the mean kinetic energy of stationary motions. It seems quite natural to extend this assimilation to all substances. Also, right from the beginnings of the Mechanical Theory of Heat, had not Clausius and Rankine hesitated to regard this assimilation as legitimate. The proposition to be proved may then, in algebraic language be stated thus: *The average kinetic energy of the stationary motions is the integrant divisor of the quantity of heat released.* Such is the theorem that Boltzmann in 1866 and Clausius in 1871 strove to justify.

When it is a question of interpreting the First Principle of Thermodynamics, one can leave a very large indetermination[1] in the

---

[1] *Translator's Note:* The term 'indetermination' has been chosen for two particular reasons. Firstly, since the notion of quantum

nature of the stationary motions that animate the atoms. To prove the theorem we have just enunciated, neither Boltzmann nor Clausius could preserve such an indetermination; they had to adopt more restricted hypotheses; they assumed that each of the atoms of a body in apparent equilibrium travels over a closed, or very nearly closed, trajectory, and that these atoms describe their orbit in the same time; they admitted that the forces acting on each atom depend exclusively upon the position of that atom, which would happen if they originated from immobile centres, but which cannot happen if they result from reciprocal actions of atoms in motion. These restrictions exclude systems whose points move in all directions, in a random walk; they also exclude systems whose particles act upon each other; hence they reject perfect gases such as Clausius's and Maxwell's kinetic theories imagined; as a result they greatly diminish the interest offered by the analysis of Boltzmann and Clausius.

Another, graver, objection arises against this analysis. In the domain of Pure Thermodynamics, a system can only be in equilibrium if it has the same temperature at all its points; hence, if one wishes to obtain a new system in equilibrium by the union of two systems in equilibrium, it will be necessary for the two systems coupled together to have the same temperature.

Let us translate this proposition of Thermodynamics into the language of the Mechanical Theory of Heat, and do so adopting the assumptions of Clausius and Boltzmann; it will take the following form: In order that the union of two systems in statistical equilibrium give a new system in statistical equilibrium, it

---

mechanical indeterminacy, now familiar, and almost automatically called to mind whenever the word 'indeterminacy' is used in a scientific context, as here, had not been proposed, and certainly not been implied by any mathematical relationship in any theory at the time, the use of the word 'indeterminacy' is completely inappropriate; its use would unjustifiably cause to intrude upon a reader's consideration of this work a set of notions completely alien to it. Secondly, to use the word 'indeterminateness', although such might at first sight, and illusorily, appear correct, would again conjour too many tacit connotations of indeterminacy, in the quantum mechanical sense, in the reader's mind. Its use is open to the sharp criticism that this word could cause the reader to imagine that at the historical moments being discussed by Duhem there were already some thoughts in the back of investigators' minds that were, unwittingly, the precursors of quantum mechanical indeterminacy. Such was emphatically not the case.

'Determination' in these contexts refers to apprehension by the vehicle of experimental procedures of measurement, the result of which is to 'determine' the values of certain measurable quantities. Since experimental procedures are subject to finite degrees of error, such determinations are imprecise; they therefore result in a degree of 'indetermination'. It is this sense that is used. It is also Duhem's word.

is necessary that the first two systems have stationary motions with the same mean kinetic energy. If the mean kinetic energy can be taken legitimately as the measure of the absolute temperature, this principle must flow from the principles of Mechanics and some hypotheses made about the stationary motion which constitutes heat. Now this essential proposition was not only not proved by Clausius and Boltzmann, but one does not even begin to see a proper method of obtaining it from their formulae.

This difficulty, the solution of which was not allowed to be either derived nor suspected, doubtless contributed to turn geometers away from attempts with the need of linking the Theory of Heat with Dynamics. Many of them, having left the principles of Thermodynamics unexplained, contented themselves with applying these principles with an ever increasing success to the diverse problems of Physics. In fact, we saw the mechanical explication of Carnot's Principle was somewhat delayed until 1884, the time when Helmholtz, in his turn, attempted it.

Helmholtz, it is true, no longer approached the problem with the great hopes and immense thought that motivated Boltzmann and Clausius; for him it was no longer a question of deducing all the laws of Thermodynamics from the principles of Dynamics, alone, applied to a certain stationary motion, and of presenting this reduction as the *mechanical explication* of effects analysed by the Theory of Heat; it was simply a matter of discovering, in the study of *monocyclic systems*, certain simple mechanisms whose motion is governed by equations *analogous* to the thermodynamical relations. Let us leave Helmholtz himself to define for us the object of his investigations[1]: "I write with the object of proving that there exist motions whose mechanical nature is entirely accessible to our understanding and in which the transformation of work into its equivalents is subject to conditions similar to those the Second Principle imposes upon calorific motion. Calorific motion is presented to us, in the first place, as a motion of an unknown kind; if one excepts the one case treated by the Kinetic Theory of Gases, the hypotheses that it has been possible to make on this subject up to now are extremely vague. In such a state of affairs I have deemed the following method to be quite natural: Take the most general properties of calorific motion that are known to us and investigate under which very broad conditions these properties are found in the other well known classes of motions. My researched in this direction have led me to discover the analogies that exist between calorific motion and the monocyclic motions I have studied. But I have constantly brought into evidence this truth, which I had stated at the beginning: Rigorously speaking, calorific motion cannot be monocy-

---

[1] H. von Helmholtz: Studien zur Statik monocycklischer Systeme, (Zweite Fortsetzung), *Sitzungsberichte der Berliner Akademie*, (10th July, 1884), p. 757; *Wissenschaftliche Abhandlungen*, vol. III, p. 176.

clic. Also, I have never had any pretension of giving an *explication* of the Second Principle of Thermodynamics."

Boltzmann, expounding Helmholtz's theories in a more precise form still, expresses it in this way[1]: "These theories rest upon some hypotheses that have the aspiration of expressing the very constitution of the primordial elements and primitive forces of Nature; they treat simply certain mechanisms whose behaviour presents, in one respect or another, a considerable analogy with the interplay of natural phenomena. The more impressive this analogy is, the more numerous are the particularities it reproduces, the more useful is the mechanism employed. According to Maxwell's word for it, this mechanism is a dynamical illustration."

Let us give a succinct ideal of the theory of monocyclic systems.

Observe a top *standing still*; it seems to be immobile; in reality this is not so at all; it possesses an extremely rapid rotational motion; at every instant, each of the elementary masses that compose it leaves the position that it occupies in space to go to occupy another; but immediately it is replaced by a similar mass, so that the eye perceives no alteration. This top which stands still provides us with a picture of what in Mechanics is called a system in *perpetual motion*, what Helmholtz calls a *monocyclic system in equilibrium*. We have already caught a glimpse of an analogy between such a system and those the Mechanical Theory of Heat studies: the equilibrium we observe is an apparent equilibrium, a statistical equilibrium; under this equilibrium there are hidden very rapid stationary motions.

The stationary motions constituted by the rotations of our top correspond to a considerable kinetic energy, which appears here as the internal energy; this figure for the energy would have to be completed by the addition of the internal potential if some forces were exercised between the various parts of the system.

Instead of assuming that the top keeps an unchanging position in space, we can imagine that it is displaced slowly, that its axis changes position and direction; its real motion is then composed of two kinds of motion: a very rapid rotational motion that produces no apparent alteration in position, and a motion very slow with respect to the former; this motion alone is perceptible; whilst the first embodies the stationary motions whose existence is postulated by the Mechanical Theory of Heat, the second represents the observable alterations in the state.

Let us imagine that an external action intervenes to produce one of these alterations, it inclines slowly the axis of the top, it modifies the disposition of some one of its parts. The work that this action performs in producing this perceptible alteration is not equal to the work it would have performed in modifying in the same way the position or shape of top deprived of all

---

[1] L. Boltzmann: *Vorlesungen über Maxwell's Theorie der Elektricität und des Lichtes*, Part I, (Leipzig), (1891), p. 13.

rotational motion; this latter work is only a part of the former; here it represents what the Mechanical Theory of Heat calles the *external work*. But another part of the work performed by the external actions has had no visible employment; it has contended against the inertial forces arising from the rotational motion of the top; it has modified the kinetic energy of this motion; to follow our analogy, we shall say that it represents the *quantity of heat absorbed* by the system.

Upon analysing the motion of a monocyclic system such as our top, we recognise some quantities suitable for representing the internal energy, the external work, and the quantity of heat given off; furthermore, it suffices to appeal to the dynamical law of kinetic energy to obtain these quantities between a relation similar to the equation for the equivalence of heat and work. Could one equally well make them enter into a relation analogous to that given by the Carnot-Clausius Principle? Taking the ratio of the quantity of heat released during an elementary modification to a suitable integrant divisor, can one equate this quotient with the decrease suffered by a certain function that plays the rôle of Entropy?

One can prove the existence of such an integrant factor on the condition of restricting the generality of the monocyclic systems studied; unfortunately it is difficult to interpret, within the meaning of the Mechanical Theory of Heat, the restrictive conditions to which one must appeal. Tightening the conditions further, one can even make this integrant divisor the kinetic energy of the stationary motions, and obtain from that a more intimate agreement between the statics of monocyclic systems and the Mechanical Theory of Heat of Boltzmann and Clausius.

Here we again find a question that has already attracted our attention.

In order that the union of two thermodynamical systems in equilibrium provide a system in equilibrium, it is necessary for the two component systems to have the same temperature; this common temperature is then that of the resultant system. If we wish to find some monocyclic systems whose properties can *illustrate* the thermodynamical equations; if we wish, in particular, that the integrant divisor of the quantity of heat released be the *mechanical model* of absolute temperature, these monocyclic systems will have to satisfy the following proposition, here stated: By suitably uniting two monocyclic systems with the same integrant divisor, one obtains a new monocyclic system which admits as integrant divisor the common integrant divisor of the first two.

The study of this *isomoric coupling* (ἴσον μόριον, equal denominator) had occupied Helmholtz for a long time; he had given the analytic expression to the conditions outside of which isomoric coupling could not take place; but it is quite difficult to comprehend a reconciliation between these conditions and the hypotheses of the Mechanical Theory of Heat.

Thus, to define the monocyclic systems whose properties are capable of imitating the thermodynamical relations, Helmholtz was obliged to submit them to conditions that expressed certain analytic characteristics of the functions used; it is quite difficult to translate them into mechanical language, and even more difficult to draw from them some precise information about the assumptions that it would be useful to make, touching upon the structure of atoms or the nature of calorific motion. From there it is permissible to ask if this analogy between the laws of monocyclic systems and the equations of Thermodynamics has any foundation in the nature of things.

Between the equations of Thermodynamics and the mechanical properties of the systems studied by J. Willard Gibbs[1], the analogy is certainly more direct and capable of being taken further. The hypotheses that serve as the point of departure for Gibb's researches are a kind of generalisation of those that have served as a basis for the Kinetic Theory of Gases; these hypotheses are developed with a rigour and clarity that are admirable.

In a certain space an immense number of bodies is distributed, varying in shape and position. All these bodies, which are the elements of the system studied, are of the same nature; they could be taken to a stage, where they would all be identical; but at the moment when we study them they differ from each other in their state, for they are variously located, oriented and de-deformed, and in their motion, for they do not all have the same velocities. A large indetermination is left in the nature of these bodies. They may be simple point masses; the position of each of them depends solely, then, upon three coordinates. They may be rigid atoms; to know the position of such an atom, it is necessary to know the values of six variables. They may be molecules, assemblies of atoms more or less numerous, the more or the less different, capable of being displaced with respect to each other; to determine the shape and position of such an assembly it is necessary to be given a smaller or larger number of variables, but greater than six. One condition alone is required of the elements which form the mass system studied; in this way such an element is entirely known in shape and position when one knows the values of a larger or smaller, but limited, number of independent variables.

These elements are subject to certain forces. The forces which act upon an element depend exclusively upon the variables which determine this element; such would be forces originating from unchanging external bodies. Such a hypothesis evidently excludes the hypothesis of reciprocal actions between the elements; as one has no longer assumed these elements to be capable of coming into collision, Gibbs' theory rejects as outside its domain the various forms of the Kinetic Theory of Gases proposed by Clausius and Maxwell. From this it reconciled the attempts made by Boltzmann and Clausius to reduce Carnot's Principle to a mechanism.

Let us assume the *statistical equilibrium* of the system to be

---

[1] J. Willard Gibbs: *Elementary Principles in Statistical Mechanics*, (New York and London), (1902).

established. A crowd of distinct states with distinct motions are simultaneously realised; at each instant, each of the elements leaves its state and its motion; but another element, at the same instant, palpably takes the state and motion that the former has just lost.

How are all these states and all these motions distributed amongst all the innumerable bodies that form the system? At a given instant how many bodies are there whose state lies between two given limits, whose motion also lies between two given limits? Such is the first problem a geometer[1] has to set himself. It is analogous to this other one, familiar to actuaries: In a country with a population that is stable and which has a determined number of inhabitants, how many men are there whose age lies between two given limits? The methods of the Calculus of Probabilities draw the solution of the latter problem from mortality tables; they draw the solution of the former from the Principles of Mechanics. Maxwell and Boltzmann had already given this solution under the circumstances where the Kinetic Theory of Gases holds; Gibbs developed it for the very general systems that he proposed to study.

The distribution law for the various states and various motions at the heart of a system in statistical equilibrium is subject only to very broad conditions; amongst all the forms, finite in number, in which it may be given, there is one in which it appears as being endowed with particularly simple algebraic properties. Gibbs named this distribution law the *canonical distribution*. The distribution law that Maxwell's theorem imposes upon the velocities with which the atoms of the gas move is a very particular case of a canonical distribution.

Gibbs takes the systems with a canonical distribution as the proper object of his analysis. In the formula that governs a canonical distribution, there occurs a certain quantity, the *distribution modulus*, which was going to play an essential rôle in thermodynamical analogies; it is the distribution modulus which is to represent absolute temperature in these analogies. In the particular case where the bodies which form the system are reduced to point masses, the canonical distribution law reduces to that which Maxwell announced; the distribution parameter is then identical with the mean kinetic energy; hence if one simply wish-

---

[1] *Translator's Note:* It was only in the last half of the nineteenth century that the widespread development of Analysis took place; consequently students of today are frequently familiar with much that was foreign or very novel to mathematicians of that period. Duhem, himself, was naturally abreast of developments, and so referred to mathematicians of earlier times as 'geometers', for such was the foundation of their mathematical interest, by and large. It would be significantly misleading to use the name 'mathematician' where Duhem uses the name 'geometer', for it would tacitly imply the possession of knowledge by those investigators referred to by Duhem that they could never have possessed.

ed to compare the bodies studied by Thermodynamics with systems of free point masses, one would have to take the mean kinetic energy of the molecular motion as the measure of absolute temperature; this, in fact, is what Boltzmann and Clausius did. But if the molecules do not reduce to simple point masses, if they are complicated, the mean kinetic energy will no longer be the parameter of the canonical distribution, it will no longer represent absolute temperature.

The analogy between the distribution modulus and absolute temperature is affirmed, first of all, by the following propositions which nicely display the superiority of Gibbs' analysis over the attempts of his predecessors:

When one couples two systems in statistical equilibrium, both endowed with a canonical distribution, the resultant system can only be in statistical equilibrium if two component systems have the same distribution modulus; the resultant system then admits a canonical distribution with the same modulus as the component systems. If the two component systems do not have the same distribution modulus, their coupling breaks their state of equilibrium and obliges them both to be modified; the one admitting the greater distribution modulus loses energy, the other gains.

Nevertheless, the equations which control our system in statistical equilibrium are not absolutely similar to the thermodynamical formulae; the variations depend upon the number of variables that have to be known to determine the shape and position of each of the elements of the system; these variations are all the smaller the larger is the number of variables; one can therefore give the following conclusion to Gibbs' researches: The equations of Thermodynamics represent the limiting form of the laws which control the statistical equilibrium of a system with a canonical distribution when one increases beyond every limit the number of variables necessary to define each of the elements of the set.

This conclusion of Gibbs' investigations is very unexpected. It shows that physicists desirous of explicating phenomena by 'mechanical arguments' must renounce hypotheses that give to atoms a very simple constitution, such as point masses or rigid solids; amongst the properties of mechanisms which were imagined and the natural laws, they could only hope for an approximate agreement by assimilating atoms into extremely complicated assemblages; if they desire not an approximate agreeement, but a rigorous one, they would have to conceive of atoms with an unlimited number of variables, of continuous and deformable small bodies, such as small fluid masses would be; the consideration of fluid atoms would take us a very long way from the principles so dear to atomists.

The theory of J. Willard Gibbs is assuredly the most powerful attempt that has been made up to now to reduce the laws of Thermodynamics to the Principles of Mechanics; however, it falls short of having pushed this reduction to the point where there is nothing else to be desired; more than one question arises naturally, which up to now have had no answer. Here is the first:

The ensembles with a canonical distributron are defined by a purely algebraic characteristic, by the form of the equation which governs the distribution of the various states and the various motions in the body of the system in statistical equilibrium. Is it possible to bring a mechanical characteristic into correspondence with the algebraic characteristic? Can one say how the elementary bodies forming an ensemble have to be constituted, to which forces they must be subject, in order that this ensemble in statistical equilibrium can be assigned a canonical distribution?

This question is still without an answer; however, it would be necessary for it to be solved before one could attempt to answer this second question:

If the ensembles with a canonical distribution have attracted the attention of geometers, it is solely because their algebraic study was declared to be particularly simple and easy. For what reason did the systems studied in Thermodynamics become reconciled with ensembles with canoncial distributions rather than other ensembles? The properties of an ensemble in statistical equilibrium, but where the distribution is not canonical, doubtless differ very much from the laws of Thermodynamics; how is it that nature presents us with no system endowed with such properties?

As long as this question receives no satisfying answer it will be difficult to regard the *mechanical explication* of the Principles of Thermodynamics as complete. In every case this explication seems still far distant; all that is logically permissible to affirm is that it is, nevertheless, possible to construct mechanically, at least to define by certain algebraic conditions, some ensembles of bodies whose stationary motions are governed by formulae analogous to the equations of Thermodynamics. To go back to a few words which Boltzmann borrowed from Maxwell, *the Mechanical Theory of Heat does not furnish a mechanical explication of Thermodynamics; it gives only a dynamical illustration of it.*

CHAPTER XI

THE MECHANICAL THEORIES OF ELECTRICITY

The attempts at explaining electrical phenomena mechanically are innumerable; the study of these attempts suggests some reflections similar to those that one can draw from the mechanical theories of heat; it is these reflections which are important to our object rather than the very detail of the explications; we shall not undertake, therefore, to pass them all in review and we shall keep only to those which are most in vogue, to the theories of Maxwell.

We owe to Maxwell two attempts, which led by very different methods to the mechanical explication of electrical phenomena. The first in date is that expounded in the memoir entitled: *On Physical Lines of Force*; it consists in imagining completely a mechanism capable of explaining electrostatic and electromagnetic effects.

Maxwell imagined a non-conducting body — in this attempt he considered none other — in the image of a honeycomb; the walls of wax are replaced by partitions that form an isotropic, perfectly elastic solid; the honey is described by a perfect fluid that is set in motion with extremely rapid vortex motions; the deformations which the elastic walls suffer, the pressures and tensions that these deformations generate, explain the phenomena that we attribute to the polarisation of dielectrics; the vortex motions of the intracellular liquid, the inertial forces which result from them, given an explanation of the effects that we attribute to magnetisation.

We shall not tarry to discuss here the insufficiencies of this explication, the errors of calculation or argument which Maxwell has dotted about, the incompatibilities between the results obtained and very certain laws of electricity and magnetism; we have detailed this discussion elsewhere[1]. Doubtless, therefore,

---

[1] P. Duhem: *Les théories électriques de J. Clerk Maxwell; Essai historique et critique*, (Paris), (1902).

Maxwell found little satisfaction in the mechanism he had thought of, for he soon abandoned it to set out upon a completely different path towards the mechanical explication of electric phenomena[1]. Here are the terms in which he himself defines this new method[2]:

"In this Treatise I propose to describe the most important phenomena by showing how one can submit them to measurement and investigate the mathematical relations which exist between the quantities measured. Having thus obtained the data of a mathematical theory of Electromagnetism, and having shown how this theory can be applied to the calculation of phenomena, I shall strive to bring to light as clearly as I can, the relations which exist between the mathematical forms of this theory and those of the fundamental science of Dynamics; in that way we shall be, in a certain measure, prepared for defining the nature of the dynamical phenomena amongst which we must look for analogies or explications of electromagnetic phenomena."

How Maxwell intended to follow the method he had just defined is what we are now going to examine[3].

Let us refer to what was said earlier in the book about Lagrange's Analytical Mechanics, and let us recall in what way it forms the equations of motion of a system.

To represent the state of this system it uses a certain number of independent variables $\alpha, \beta, \ldots$; the first derivatives of these variables with respect to time are the *generalised velocities*; their second derivatives are the *generalised accelerations*.

Once the independent variables are chosen, it has only to consider three mathematical expressions which, by regular calculations, furnish it with the equations it is wished to be obtained. These three expressions are:

(1): *The virtual work of the external forces*: the knowledge of this work is equivalent to the knowledge of the generalised exterior forces which correspond to the various independent variables; if the state of the exterior bodies is given, these generalised forces depend only upon the variables which fix the state of the system and not upon the generalised velocities nor the generalised accelerations.

(2): *The internal potential*: this is a quantity entirely defined by the knowledge of the independent variables, without any intervention of velocities or generalised accelerations.

---

[1] J. Clerk Maxwell: A Dynamical Theory of the Electromagnetic Field, *London Philsophical Transactions*, vol. CLV, (1864); *Scientific Papers*, vol. I, p. 526; *Traité d'Électricité et de Magnétisme*, (translated into French by G. Seligman-Lui), Part IV, chapters V,VI,VIII; vol. II, pp. 228-262.

[2] J. Clerk Maxwell: *Treatise on Electricity and Magnetism*, preface to the first edition.

[3] On this subject, see: H. Poincare: *Electricite et Optique*, (1st edn), vol. I, (Paris), (1890), Introduction; (2nd edn), (Paris), (1901), Introduction.

(3): *The kinetic energy*: this last quantity no longer depends only upon the independent variables, but also upon the generalised velocities; with respect to the latter it is homogeneous and of second degree; finally, it can only be zero or positive.

What course will we have to follow if we want to prove that a set of phenomena, for example the set of electromagnetic phenomena, is susceptible of a mechanical explication?

First of all we shall admit that the experimental method has represented by measurable quantities all the properties which are displayed in the phenomena studied, that it has expressed all the laws which these phenomena obey in the form of equations between these various quantities.

Then taking the set of measurable quantities by which are represented the properties of the system studied, we shall separate them into two categories: one will be regarded as independent variables; the other will be generalised velocities corresponding to the variables of which we have just spoken or to other variables which are not directly revealed to an experimenter.

Thus the quantities which fix in space the position of the various bodies, the components of the dielectric polarisation on each of them will be regarded as independent variables; the velocities of the perceptible motions correspond to the first variables; the generalised velocities which correspond to the second variables are what Maxwell calls the *components of the displacement flux*; without being precisely generalised velocities, the *components of the conduction flux* are linked to the velocities with which the electric densities vary.

By means of these various quantities we shall form two combinations: one which will be treated as *internal potential*, the other as *kinetic energy*; the first will have to contain only independent variables and no generalised velocities; the second will not contain variables alone, but also generalised velocities; with respect to the latter it will be homogeneous and of second degree; finally, it will be negative.

For example, we shall never count the *electrostatic potential* as being part of the internal potential. The *electrodynamic potential* depends upon the intensities of the conduction currents and displacement, intensities which we shall regard as generalised velocities or as related to these velocities; it is homogeneous and of second degree in the intensities; finally, it is never positive; we shall subtract it from the kinetic energy of the perceptible motions to have the total kinetic energy.

Let us give the virtual work of the external actions to which the system is subject, and we shall be provided with all Lagrange's method requires to form regularly the equations of motion of our system. So let us form these equations; if they are identical with those that the inductive method has drawn from experiment, to those which express the laws of Coulomb, Ampère, Faraday, Lenz, Neumann and Weber, we shall have proved that electromagnetic phenomena are capable of a mechanical explicat on.

Such was the method thought of and followed by Maxwell.

The explication of electromagnetic phenomena sketched thus[1], ran into grave objections; it met them particularly in studying systems which contain magnets.

Returning to the analogy that Ampère had brought to light, Maxwell assimilated each magnetic element with a small closed current; the intensity of magnetisation is then a combination of generalised velocities; it does not appear in the internal potential, but in the kinetic energy. Unfortunately this opinion attributes an unacceptable form to the internal energy of a system where there are magnets; its consequences are irreconcilable with the calorific effects produced in a mass of soft iron that a current magnetises or demagnetises.

One can avoid this difficulty by regarding the components of magnetisation no longer as combinations of generalised velocities, but as independent variables which represent a state of displacement or deformation of a certain medium; they are then analogous to the components of the dielectric polarisation, and the magnetic potential appears in the internal potential in the same capacity as the electrostatic potential. But if this be so, the velocities with which the components of magnetisation vary would have to appear in the expression for the kinetic energy as the components of the displacement flux appear; the presence of these velocities in the kinetic energy would have to give rise to inertial forces analogous to the electrodynamic forces; now, no experiment has, up to now, revealed the actions produced by such *magnetic displacement currents*.

Let us condemn these objections. Let us argue as if Maxwell's analysis were faultless.

When we defined an internal potential and a kinetic energy, when, by Lagrange's method, we found them from equations which agree with the experimental laws of a group of phenomena, does it follow that this group of phenomena is mechanically explicated? We have evidently satisfied necessary conditions for this group of phenomena to be mechanically explicatable; but are these conditions sufficient? From the internal potential containing only the independent variables, from the kinetic energy being homogeneous and of second degree in the generalised velocities, from it never being negative, can we conclude with certainty that there exists a certain grouping of masses and forces, a certain mechanism, admitting such a potential and, above all, such a kinetic energy? Cannot the form of the latter exclude the possibility of such a mechanism under certain circumstances? Thus, in the case treated by Maxwell, the system is the seat of three kinds of motions: the perceptible motions, the stationary motions which constitute heat, and the motions which are manifested to us by electric currents; it has been assumed that the kinetic energy

---

[1] One will find a very clear and very concise exposition of this method in: E. Sarrau: Sur l'application des équations de Lagrange aux phénomènes électrodynamiques et électromagnétiques, *Comptes rendus*, vol. CXXXIII, (1901), p. 421.

of the system is the sum of the kinetic energies of each of these three kinds of motion; is it certain that one can actually construct a mechanism having these three motions and whose kinetic energy enjoys such a property?

It seems imprudent to dismiss similar difficulties with a stroke of the pen. What has been found to be best, up to now, for clearing objections of this nature, is to imagine extremely simple mechanisms whose internal potential and kinetic energy offer, in their various particularities, as more or less direct analogy with the potential and kinetic energy that it is proposed to study; in a word, this is to construct *models* which imitate in their laws of motion the equations that are discussed upon. Aided by the theory of monocyclic systems, Boltzmann[1] has illustrated the views of Maxwell on the analogy between Lagrange's equations and the Laws of Electrodynamics with such models.

---

[1] L. Boltzmann: *Vorlesungen über Maxwell's Theorie der Elektricitat und des Lichtes*, Part I, (Leipzig), (1891).

CHAPTER XII

THE IMPOSSIBILITY OF PERPETUAL MOTION

Let us forget the objection we have just raised; let us regard it as null and void; let us allow that a group of phenomena will be mechanically explicated when there is defined an internal potential and a kinetic energy from which can be obtained, by Lagrange's method, equations conforming with the experimental laws of the phenomena. The question that we shall then have for examination is the following: Can all the laws physicists establish by the inferential method be put into the form of Lagrange's equations?

An observation somewhat less careful of physical phenomena seems to authorise the following conclusion: *There exists a radical incompatibility between Lagrange's Mechanics and the Laws of Physics; this incompatibility does not just affect the laws of phenomena whose reduction to motion is the object of the hypothesis, but also the laws that govern perceptible motions.*

Let us bring this incompatibility into evidence by some very simple examples.

The most immediate consequence of Lagrange's equations is assuredly the kinetic energy equation. If the forces which act upon a system depend upon a potential, the sum of this potential and the kinetic energy remain constant during the whole duration of the motion of the system. Now the reciprocal actions of the various parts of the system always depend upon a potential; hence it is sufficient that the external forces depend upon a potential in order for the system to be subject to the law whose enunciation we have just recalled; in particular, this theorem is applicable to a system which suffers a single external action, that of weight.

Let us follow such a system in its motion; each time that it takes the same shape and passes through the same position, the potential of the forces as much internal as external takes the same value again; the kinetic energy therefore also retakes the same value.

This *conservation of kinetic energy* is one of the most obvious consequences of the Dynamics of d'Alembert and Lagrange; does it accord with the lessons of experiment, I mean the most basic experimentation?

Here is a decanter full of water. I shake it vigorously and I put it on the table. The water occupies a certain position and has a certain shape, that is to say the position and shape of the decanter which encloses it; this water whirls rapidly so that its kinetic energy has a considerable positive value. At the end of a quarter of an hour the water still has the same shape and position; according to Lagrange's Mechanics it would have conserved its kinetic energy; but it is now at rest and its kinetic energy is zero.

A plumb line hangs vertically. By a sharp blow I give it an initial velocity and, consequently, an initial kinetic energy. I let it oscillate, and at the end of some time I observe it anew; it hangs vertically; the potential of the weight, which acts upon it, has the same value as at the beginning of the motion; the same will have to hold for the kinetic energy; none at all; the plumb line is now immobile and the kinetic energy is zero.

Thus the simplest observations show us that natural motions contradict the law of the conservation of kinetic energy.

The analysis of the motions of our plumb line will allow us to sharpen the form of the disagreement between the Lagrange equations and the natural motions; to this end let us stop for an instant to consider the constitution of Lagrange's equations.

A system is assumed to be subject to the action of external bodies which stay unchanged throughout the whole duration of the motion. According to the Principles of Dynamics:

(1): The generalised external forces depend exclusively upon the variables which determine the state of the system;

(2): The internal potential and, therefore, the generalised internal forces depend exclusively upon the same variables;

(3): The kinetic energy depends upon these variables and the generalised velocities; it is homogeneous and of second degree in these velocities.

From these Lagrange's process for calculating the generalised inertial forces teaches us that each of these forces is a sum of two terms; that these two terms contain the independent variables; that the first is homogeneous and of second degree in the generalised velocities, but does not contain the generalised accelerations; finally, that the second, independent of the generalised velocities, is homogeneous and of first degree in the generalised accelerations.

To obtain the equations of motion, one forms, with respect to each of the independent variables, the sum of the three external, internal, and inertial, generalised forces, and equates this sum to zero. Consequently, the first member of each of these equations is a sum of three terms which each contain the independent variables; the first terms depends neither upon the generalised velocities nor the generalised accelerations; the second, inde-

IMPOSSIBILITY OF PERPETUAL MOTION

pendent of the generalised accelerations, is homogeneous and of second degree in the generalised velocities; the third, independent of the generalised velocities, is homogeneous and of first degree in the generalised accelerations.

This composition of Lagrange's equations implies a consequence to which we are going to give a precise form.

Let us assume that these equations are satisfied when the system is in a certain state, when its various point masses are provided with certain velocities and certain accelerations; they will still be satisfied if one takes the system in the same state with the same accelerations and if one reverses the directions of all the velocities without altering their magnitude. This proposition, which clearly flows from the preceding can be stated again in the following way: Lagrange's equations are satisfied by a motion which makes a given sequence of states traverse the system; they will be satisfied by a motion which makes the system pass through the same states taken in reverse order, and such that the interval which separates the two given states is always crossed in the same time in the course of the two motions.

From this proposition it is not difficult to draw this conclusion following:

Let us assume that the system, starting from a certain initial state A with certain initial velocities V, arrives under the action of certain forces at a certain final state Ω with certain final velocities V'. Let us place it in the state Ω with velocities equal and directly opposite to the velocities V', and then subject it to the same forces; it will arrive at the state A with velocities equal and opposite to the velocities V; and the two motions will last for the same time.

Such is the essential characteristic, which we can summarise in these words: *All motions controlled by the Dynamics of d'Alembert and Lagrange are reversible motions.*

Now let us go back to our plumb line. We pull it aside by a certain angle to the left of the vertical, so taking it to a position A, and then we leave it on its own; it returns to the vertical, passes it, and on the right reaches an extreme position Ω where the velocities of all its points vanish. By virtue of the previous proposition, it must now take the reverse motion, return to the position A and recommence indefinitely these oscillations unvarying in amplitude and duration. This is not what takes place. Having left the position Ω, the pendulum regains the vertical and passes it; but it stops before having reached the position A; the successive oscillations thus continue with decreasing amplitude and reducing the line little by little to its equilibrium position. This example shows us that the *natural motions are not reversible*.

If the equations of Dynamics given by Lagrange represent reversible motions exclusively, which they must be in the absence of every term of odd degree in the generalised velocities, they will therefore be able to be made to lose this characteristic, and one will obtain equations which represent non-reversible motions if one introduces terms of first degree in the velocities. For that it will suffice to submit the system not only to the forces which we have considered up to now and which depend only upon the positions of its various parts, but also to some forces that depend upon the velocities with which these parts move, provided that these forces change in direction when all the velocities are reversed.

Thus the damped oscillations of our plumb line will be very accurately represented by assuming that the motion of this pendulum suffers a resistance proportional to the angular velocity; thus, further, Navier had been able to give to the equations of Hydrodynamics a form excluding reversible motions and conservation of kinetic energy by assuming that the fluid's molecules exercise upon each other reciprocal forces that depend upon their relative velocities.

From the point of view of Algebra, this generalisation of the equations of Dynamics was easy to notice; moreover, Lagrange had indicated it[1]. But from the point of view of Physics it constitutes a profound transformation of the hypotheses upon which the science of motion rest, an upheaval of d'Alembert's Principle. The statement of this principle only has a meaning if the real forces to which a mechanical system is subject remain the same for one state of the system, that the system is at rest in this state or which it passes through in the course of a motion. If the actual forces were changed so that instead of conceiving of a system in motion in a certain state it is assumed to be at rest, one would formulate a nonsense in stating d'Alembert's Principle: A system in motion can be maintained in equilibrium in each of the states it passes through if the inertial forces are adjoined to the actual forces which act upon it when it is in this state.

Do we have to conclude this discussion about the existence of an essential incompatibility between the natural motions of Dynamics drawn from d'Alembert's Principle by having to modify the latter profoundly? The incompatibility, Helmholtz showed, can definitely only be apparent. Let us imagine that in a mechanism there are masses with motions that our senses cannot perceive. Although the real and complete laws of motion of this system are given by the equations of Lagrange's Dynamics, it can certainly happen that the laws experimentally established, *and which are incomplete*, seem to contradict this Mechanics; in particular it may happen that the observable motions appear as non-reversible.

To explain Helmholtz's thought, let us analyse the example he himself chose[2].

If Lagrange's equations can only represent reversible motions,

---

[1] Lagrange: *Mécanique analytique*, (2nd edn), Part II, Section II, n° 8.

[2] H. von Helmholtz: Studien zur Statik monocycklischer Systeme I, *Sitzungsberichte der Berliner Akademie*, (6th March, 1884), p. 169; *Borchardt's Journal*, vol. XCVII, p. 121; *Wissenschaftliche Abhandlungen*, vol. III, p. 131.

they have to do so with the absence, in their composition, of every term of odd degree in the velocities; this absence itself arises from this, that the kinetic energy contains only terms of second degree in the velocities.

Let us imagine a body which turns around a vertical axis; its kinetic energy is obtained by taking half of the product of its moment of inertia by the square of the angular velocity of rotation.

Let us assume that this body carries a centrifugal force regulator mounted on the same axis. During a period of variable angular velocity of rotation the arms of the regulator are deflected, and the motion of the system is no longer a simple rotational motion; the kinetic energy has an expression more complicated than that of which we have just spoken. Once a permanent régime is established, the balls on the regulator keep at a definite deflection; the kinetic energy is obtained by multiplying half the square of the angular velocity by the sum of the moments of inertia of the body and of the moment of inertia of the regulator. The first moment of inertia is fixed, but the second changes with the angular velocity of rotation, so that even in a permanent régime the kinetic energy is no longer simply proportional to the square of the angular velocity. Let us imagine, for example, a combined regulator of such a kind that its moment of inertia in the permanent régime varies proportionally to the angular velocity of rotation; the kinetic energy of the mechanism provided with a uniform rotational motion will be the sum of two terms, one proportional to the square of the angular velocity, and the other to the cube of this velocity; during a variable period a third term will be added to the latter two; in all circumstances the kinetic energy will keep a term of odd degree in the angular velocity.

Let us now consider the regulator to be made of such a substance that we can perceive neither its existence nor its movements; the *experimental* study of the rotational motion of the body would show us that its kinetic energy includes a term proportional to the cube of the angular velocity; Lagrange's Dynamics would seem to be contradicted by this study; it would be confirmed if we could take account of the *hidden motions* of the regulator.

Here is another experiment from the domain of entertaining Physics which brings out fully Helmholtz's idea:

Two eggs are on a plate; one has not yet been subdued by cooking, the other has been thoroughly baked in boiling water; as with teetotums let us impress upon them a very rapid rotational motion; the hard egg turns for a long time, losing very slowly the kinetic energy that was communicated to it; the raw egg is stopped almost immediately; the *hidden motions* of the yolk and the white have seemed to put to default the principle of the conservation of kinetic energy.

It will therefore be possible to re-establish the agreement between Lagrange's Dynamics and Experimental Mechanics if it is admitted that the observable motions are not the only motions that activate natural systems; that to these motions there are added *hidden motions* which escape our direct observation; that alone, the divergences of which these motions are the explication allow us to make out the particularities of them.

The very experiments by which we have brought into evidence the divergences between the natural motions and the Dynamics of d'Alembert and Lagrange are going to serve us as examples for showing the use physicists have made of hidden motions for a long time.

The oscillations of a pendulum continue slackening; physicists attribute this dampening to the movements that the pendulum communicates to the surrounding air; having adopted this explication, the experimental study of the law of dampening of the oscillations of a pendulum becomes a very sensitive means of studying certain particularities of the motion of fluids.

A fluid set in rapid motion and enclosed in an immobile receptacle returns to rest little by little. To explain this fact and several others, Navier modified d'Alembert's Principle and considered the forces of viscosity to be related to the relative velocities of the molecules. Without renouncing the Dynamics of Lagrange, and by assuming only that the gaseous molecules are point masses which are repelled in inverse proportion as the fifth power of their mutual separation, the kinetic theory established the laws of motion of gases; the perceptible motions are similar to those that Navier's hypothesis forecast; the rôle that viscosity plays in this hypothesis is taken, in Maxwell's hypothesis, by the hidden motions which violently agitate the molecules, and which our grosser senses cannot perceive.

Can all the divergences that experiment manifests between non-reversible natural motions and reversible motions foreseen by Lagrange's equations be explained by the introduction of hidden motions? It does not seem that one can answer this question negatively with certainty. Since there is imposed no condition, any restriction, upon the hidden motions, upon what basis could one prove that a given discrepancy could not find its justification in them? So it seems that at the point at which we have now arrived we may enumerate the following proposition:

Whatever may be the form of the mathematical laws to which experimental inference subjects physical phenomena, it is always permissible to pretend that these phenomena are the effects of motions, perceptible or hidden, subject to the Dynamics of Lagrange.

The mechanical explication of the Laws of Physics hence seems to escape the grips of every logical contradiction; this is not to say that it is fully satisfactory and free from blind spots. As long as, following Pascal's advice, it is content "to say on the whole: That is done by shape and motion", it triumphs over all objections without difficulty; but when it proposes "to say what is what and to construct a mechanism" it becomes weakened by a singular powerlessness. When observation reveals certain discrepancies between Lagrange's Dynamics and natural phenomena, defying every contradiction it can affirm that these discrepancies are owed to hidden motions; but if from the laws experimenally given by these discrepancies, one wishes to revert to the laws of hidden motions which produce them, one will find in his teachings no regular method certain of effecting such a passage: one is reduced to guessing it.

Amongst the empty regions that the theory of hidden motions produces, there is one with which we should particularly persist.

We have seen that natural motions are not subject to the law of the conservation of kinetic energy; they depart from it; but *they depart from it in a definite way, always the same*, and it is this characteristic that is going to fix our attention.

A liquid agitated by vorticial motions and enclosed in an immobile container returns to rest; the kinetic energy falls to zero. A plumb line set into a swinging motion ceases, at the end of a certain time, to oscillate; it has dissipated the kinetic energy it had been given. In one case, as in the other, there is a *loss* and not a *gain* of kinetic energy. All observations of this kind agree in showing that natural motions are subject to the following law:

When a system, acted upon by forces that are derived from a potential, is set off from a certain state with a certain kinetic energy and returns to the same state, it returns with a diminished kinetic energy; along the *closed cycle* traversed by the system, there has necessarily been a *loss of kinetic energy*.

According to this law one cannot construct a mechanism which, by itself, returns periodically to the same state and returns to it with the same kinetic energy or with a kinetic energy enhanced at each revolution; *perpetual motion is impossible*.

More generally, let us analyse an arbitrary motion of a system acted upon by arbitrary forces. The work of the forces applied to the system during a certain lapse of time is not, as required by Lagrange's Mechanics, equal to the increase of the kinetic energy during the same time; it always exceeds this increase. If it be so desired, in imitation of Navier, to explain this discrepancy by introducing into the equations of motion *viscosity forces*, related to the velocities of the various parts of the system, and these forces will not have to be arbitrary; their work, during an arbitrary lapse of time, will always be negative; these forces will therefore always tend to diminish the kinetic energy, to retard or stop the motion; these will always be *passive resistances*, never *active powers*.

Thus natural motions differ from the forecast Laws of Dynamics, and always in the same way. But this kind of impetus given to natural phenomena, always in the same sense, has not been met in the study of perceptible motions up to now. Is it also encountered when the bodies studied do not suffer simply changes of position, but even heatings and coolings, compressions and dilatations, meltings, vaporisations, chemical reactions, electrifications, magnetisations?

It was one of the traits of the genius of Sadi Carnot, and

perhaps the greatest, to proclaim that perpetual motion, already recognised as impossible for mechanical actions alone, is still so when one employs the influence either of heat, or electricity, and to found upon this affirmation the theory of the production of work by heat. The truth recognised by Carnot was finally given precise form by Clausius and Kelvin[1]; the first of these two savants gave it its definitive formula.

In Chapter X we enunciated the Carnot-Clausius Principle in the following form: When a system suffers a modification, the *transformation value* of this modification is equal to the loss suffered by the *entropy* of the system.

This law, as we have said, is one of two columns which sustains the entire edifice of Thermodynamics; the interpretation of this law by means of the equations of Dynamics is the essential problem of the Mechanical Theory of Heat that was the object of the efforts of Boltzmann, Clausius, Helmholtz and Gibbs.

Now when one compares this law with the modifications which Nature presents to us, one can make in this respect some observations analogous to those which suggested to us the experimental checking of the equations of Dynamics. Natural phenomena do not satisfy Clausius' inequality. The sum of the transformation value and the increase of entropy would have to be equal to zero in every case of a modification; it is not so; it has a certain non-zero value which is the *uncompensated transformation* with respect to the modification studied; and by bold and penetrating intuition, Clausius discovered this law: *The uncompensated transformation which corresponds to an arbitrary modification is always positive*.

Thus, all the modifications which are produced in the physical world are characterised not only by equations but by an inequality that is always in the same sense. This is what we had already recognised in the domain of Pure Mechanics, where bodies change position in space without undergoing any change of temperature or state; in this restricted case we had seen that the work of passive resistances was always negative; this latter inequality, furthermore, is a particular case of Clausius' inequality; in a purely local motion the uncompensated transformation is obtained by dividing the work of the passive resistances by the absolute temperature of the system and changing the sign of the quotient.

The following consequences have been drawn from Clausius' inequality:

A system completely isolated in space can neither give off heat to external bodies nor receive heat from them; every modification it undergoes has a transformation value equal to zero; the uncompensated transformation reduces to the increase in entropy; and, as the uncompensated transformation is essentially positive, one can state the following theorem:

All modifications taking place in a completely isolated system cause its entropy to increase.

---

[1] W. Thomson in the original text; he was later created Lord Kelvin.

When applied to the same system, the Principle of the Equivalence of Heat and Work also provides a remarkable proposition. A system whose isolation forbids every exchange of heat with external bodies is also screened from every external force; hence, when it is altered, the increase of internal energy, added to the increase in kinetic energy has the value zero; *every modification of an isolated system has a constant value for the sum of the internal energy and kinetic energy, a sum which we shall call the total energy of the system.*

With an audacity that no rigorous proof could justify—indeed, how can we know the limits of the Universe?—Kelvin attributed to the whole Universe the properties of a bounded system, isolated in space. Accepting this grandiose assumption, Clausius could state his two propositions which had immense repercussions:

The total energy of the Universe is constant.
The entropy of the Universe always increases.

"It is perhaps an exaggeration[1] to deduce from experimental principles, whose verifications are quite limited, general views about the future of the Universe. Let us say only that Thermodynamics authorises one to consider that the Universe proceeds fatefully in a determined direction."

This course of the Universe in a determined direction seems to escape the clutches of every mechanical explication.

Let us imagine that the attempts of Boltzmann, Clausius and Gibbs have been crowned with complete success; that, by appropriate motions subject to the Laws of Dynamics, one has taken account of all the physical phenomena in the limit where they respect Clausius' inequality; it will now be necessary to explain mechanically why this equality is constantly violated, it will be necessary to justify the existence of uncompensated transformations. To do that, the motions that imply Clausius' equality, to the *random motions* as Helmholtz and Boltzmann named them, it will be necessary to add other motions, the *ordered motions*; ordered motions with respect to random motions, will play a rôle analogous to that which hidden motions play with respect to perceptible motions in the dynamical analogies that Helmholtz imagined. As these ordered motions are left entirely arbitrary, it is permissible to assume that they may always be determined so that they generate positive uncompensated transformations, and that they agree with all observed phenomena. A formal contradiction of experiment is not to be feared by the theory that invokes them; in its boundless indeterminacy it finds an impregnable redoubt.

The difficulties lie elsewhere.

In the first place, to take account of discrepancies which exist between the actual thermodynamical facts and Clausius' equality, the theory invokes the existence of ordered motions; but it

---

[1] Émile Picard: *Exposition Universelle de 1900 à Paris. Rapports du Jury international. Deuxième partie: Sciences,* (Paris), (1901), p. 31.

prescribes no method for drawing the form of the ordered motions from the experimental laws to which these discrepancies are subject. This impression, it is true, shields the theory from experimental contradictions; but, against this, it deprives the theory of the constraint of facts.

But another point merits attention. It is no longer a question of knowing whether one can determine the hidden motions so that the work performed by the passive resistances is always negative, and the ordered motions so that they generate uncompensated transformations that are exclusively positive. It is a question of knowing if the hidden motions, left completely indeterminate, would correspond infallibly to a negative work performed by the passive resistances; if the ordered motions, whatever they may be, would necessarily give a positive value to the uncompensated transformations.

Now, the answer to these questions does not seem to be in doubt. If one leaves an unbounded indeterminacy to the hidden and ordered motions, and an unrestricted generality, nothing will fix the meanings of the discrepancies which they introduce into the equations of Dynamics, nor of the perturbations which they bring to Clausius' equality. The fictitious forces which will represent the effect of the hidden motions in Lagrange's equations, will be able to be passive resistances performing negative work; but they will also be able to be active forces, performing positive work. The uncompensated transformations resulting from the ordered motions will be able to be positive, but they will also be able to take negative values.

The following conclusion is forced upon us: Hidden motions, the ordered motions that one has to invoke to take account of the discrepancies, always in the same sense, which the actual modifications produce with respect to the Laws of Dynamics and Thermodynamics, are not completely arbitrary; they form definite category amongst the infinite diversity of possible motions.

But one is then led to wonder why, amongst the infinite variety of possible hidden and ordered motions, those alone are realised which correspond to passive resistances; why the others are never met in Nature; why, besides systems incapable of perpetual motion, one never finds systems where perpetual motion is realised. Mechanics seems to have no answer to these questions.

Thermodynamics imposes upon all phenomena of the material world a tendency in the same direction; it does not follow from this that these phenomena cannot be explained by combinations of shapes, motions, masses and forces. But the hypothesis that all the effects of raw matter are mechanical in essence takes no account of the common tendency that acts upon all these effects.

CHAPTER XIII

HERTZ'S MECHANICS

Up to now we have followed the attempts by which geometers strove to reduce all the phenomena of inanimate Nature to motions, perceptible or hidden, subject to Lagrange's equations.

Independently of purely geometrical concepts, these equations introduce a certain number of notions regarded as primary and irreducible. One can distinguish four of them which are essential; these are absolute motion, time, mass, force. These notions, foreign to Geometry, are an unbearable burden for those who would wish to see in Nature "only extension and the alteration of it alone". The former made desperate efforts to free Mechanics from this baggage of non-geometric ideas and, particularly, of the most metaphysical amongst them, the notion of *force*.

Assuredly, in respect of the actual existence of force, not all physicists share this insurmountable repugnance; there are those who admit this reality very explicitly: "The attractions that produce astronomical phenomena", said Athanasius Dupré[1], "the molecular attractions that bind them, follow laws imposed upon Nature by the all-powerful and immutable will of the Creator." Hirn, yet more formal, declared that[2], "force is neither an entity of argument nor a quality of matter, as is said so often; it exists with the same claim as matter and is a special constituent principle of the Universe."

But, if some physicists admit the actual existence of force, if, with Leibniz, they even see in it something "which is related to the soul", they are, without doubt, less numerous than those who refuse to admit the idea of force as a primary notion.

---

[1] Athanase Dupré: *Théorie mécanique de la Chaleur*, (Paris), (1869), Chapter I, p. 1.

[2] Hirn: *Théorie mécanique de la Chaleur. Conséquences philosophiques et métaphysiques de la Thermodynamique*, (Paris), (1868), p. 65.

Amongst the latter there are those, like de Saint-Venant and Kirchoff, who keep the whole of Lagrange's Mechanics but simply regard the notion of force in it to be a derived notion; who, in the product of the mass of a material point with its acceleration wish to see not a quantitative symbol capable of representing the various intensities of force, to serve as a measure of it, but the very definition of force. They have some difficulty in taking their purely nominalist doctrine logically up to physical applications, and in avoiding the more or less belated entry of the concept that they have chased away from the beginnings of Mechanics. Starting from equalities that are true *by definition*, their Dynamics unfolds with a perfect ordering and impeccable linking; but what makes for rigour also makes for sterility, for it writes only identities; to transform these identities into synthetic judgements which teach us something about bodies and their motions, it is necessary to break its analytical rigidity; at the moment of treating particular forces considered by the physicist, it is necessary for them to recover all the experimental ideas from which, in their starting points, they had stripped the general notion of force. Is this method also above all in favour in comparison with those who, after having expounded a Rational Mechanics as rigorous as infertile, on the threshold of Physics abandon their disciples ignorant of the difficulties that they are going to meet and of the methods that can solve them.

Others, with Hertz, going back to the precepts of the Cartesians and Atomists, wanted to push the explication of physical phenomena further than the reduction to Lagrange's equations; they intend only to stop in their analysis after having reduced all transformations of inanimate matter to shape, motion and mass.

It is still, however, the Mechanics of d'Alembert and Lagrange which provides them with the means of constructing an explication of the World with these elements alone.

This Dynamics, in fact, does not consider only actual forces, but, further, mathematical combinations which are homogeneous in the forces, which are measured in units of force, which play the role of forces in the equations, which are, in a word, *ficitious forces*; such are the constraint forces and inertial forces.

From there we have the consequence: When experiment manifests to us effects which to us seem to follow from real forces, it could be that we are deceived, that we have to do with apparent forces, with constraint forces resulting from the presence of a body we do not see or with inertial forces arising from a motion that we do not suspect. He who, in pulling a body to which another is connected by an invisible thread, would see the second body follow the first, would believe in a reciprocal attraction between these two bodies; he would not be deceived and would have to do business with a constraint force produced by a hidden mass. He who, ignoring the rotational motion of a gyroscope, would try to deflect the axis of the instrument and would experience an active resistance, would think that a real couple tends to keep this axis in a constant direction; he would not be deceived and would do business with an inertial force generated by a hidden motion.

According to Maxwell, since Ampère physicists were victims of an illusion of this kind when they regarded electrodynamics and electromagnetic forces as real forces. As we have seen before, that great Scottish physicist regarded these actions as inertial forces; either he imagined in the heart of cells, a fluid possessing rapid gyratory motions and to which these inertial forces would be applied; or he would draw this interpretation from inspection alone of the formulae of Electrodynamics.

In Maxwell's electric theories several of the forces that physicists regarded as real forces are therefore treated as inertial forces; certain terms, that had been brought in to account for the internal potential, were henceforward attributed to the kinetic energy; nevertheless, neither the real forces nor the internal potential were completely deleted. The elastic solid that formed the walls of the cells admitted an internal potential which varied with the deformations of these walls; thus are born the real forces which are the electrostatic forces. When Maxwell abandoned the cellular hypothesis and limited himself to giving the laws of electricity an expression that recalled Lagrange's equations, he continued to regard the electrostatic potential as representing a genuine internal potential, and not as a part of the kinetic energy.

The internal potential of the real forces which follow from it are, on the contrary, completely excluded from the construction of the aether to which Kelvin attributed the propagation of light.

In continuation of the investigations of Fresnel, Cauchy, Green and Neumann, Lamé had attributed to the aether properties similar to those of an elastic solid; this aether possessed an internal potential which depended upon deformations suffered by the medium. Now, the hypothesis of such an aether ran into grave difficulties.

In order that the small motions of such a medium might account for luminous phenomena, it was necessary that the longitudinal vibrations could not propagate, whilst the transversal vibrations propagated with the speed of light. But an elastic medium that possessed this dual property of transmitting transversal vibrations with a finite velocity and not transmitting longitudinal vibrations, was a medium the existenc of which one could not conceive, if one were to take a portion of this medium and if one were to try to keep in equilibrium by constant pressures applied to its end surface, one would only obtain a state of unstable equilibrium.

Hence, if one wished to give a mechanical explication of luminous phenomena, one would have to attribute to the charged aether propagating them a constitution very different from that geometers accorded it at the beginning of the nineteenth century.

Kelvin imagined[1] an aether absolutely different from that his

---

[1] W. Thomson: On a Gyrostatic Adynamic Constitution for "Ether", Edinburgh Royal Society, Proceedings, (17th March, 1890); *Scien-*

predecessors had conceived of. This aether was formed of small solid masses, distinct from each other, and which exercise no actual force upon the others, so that the internal potential of the medium was always to be zero. Each of these small masses turned with a high velocity around an axis passing through one of its points in the fashion of a small Foucault gyroscope; this motion generates an inertial couple that opposes an energetic resistance to every action tending to deflect the axis of rotation, whilst it does not resist any motion by which this axis would be displaced parallel to itself. The *adynamic* and *gyrostatic* aether so constituted is infinitely compressible, but it reacts against every cause that tends to impress a rotation upon any of its parts. It does not transmit longitudinal waves, whilst it transmits transversal waves with a very high, but finite, velocity, as is required by the theory of light.

The concept of adynamic and gyrostatic aether assuredly merited a profound discussion. Does this hypothesis possess the very advantages that are attributed to it. Does it avoid the objections, over stability, into which the hypothesis of the elastic aether ran? Is it not limited to passing over in silence the examination of this question of stability which, in this case, furthermore, seems hardly suitable for being tackled by a rigorous method? There would be as many problems made difficult as would be eased if we were to analyse Kelvin's theory for itself. But such is not our object; this theory is only mentioned here as something leading up to Hertz's Mechanics.

Hertz's Mechanics, indeed, is the extension to the complete physical Universe of the ideas Kelvin applied to the aether alone[1].

Hertz completely suppressed real forces from his Mechanics. The world was to be formed of bodies whose every element has a constant mass and variable velocity. For each of these elements one could therefore consider a line directed in the opposite direction to the acceleration and equal to the product of this acceleration with the mass of the element. To this quantity, by pure use of language, one can give the name of inertial force; at each instant one can also, in the usual way, form the kinetic energy of the system; from the latter expression the various inertial forces are drawn by Lagrange's formulae.

The various bodies that are considered are subject to constraints; here, as in Lagrange's Mechanics, a virtual displacement is an infinitesimal displacement which respects the constraints.

Here, then, is the fundamental postulate from whence one would be able to draw the whole of Mechanics: At each instant the inertial forces applied to an independent system are such that every

---

tific Papers, vol. III, p. 467.

[1] Heinrich Hertz: *Die Principien der Mechanik in Neuem Zusammenhange dargestellt*, (Leipzig), (1894); On the subject of Hertz's Mechanics, see: H. Poincaré: Les Idées de Hertz sur la Mécanique, *Révue générale des Sciences*, vol. VIII, (1897), p. 734.

virtual displacement imposed upon the system causes them to perform zero work.

In truth Hertz stated this postulate in an original form which seemed very different from the former; but the difference is of quite another kind, one of language, so to say; Hertz's fundamental hypothesis and the one we have just formulated are expressed by exactly the same equations.

These equations have the form given by Lagrange to the equations of Dynamics; except that there appears no real force; the forces that appear there are purely mathematical expressions, fictitious forces like inertial forces or constraint forces.

How could one, then, with these equations which no longer admit any real force, neither external nor internal, given an account of the motions that are ordinarily represented by equations of the same form, but dependent upon real forces? Wherever, in the equations of motion of a system, there appeared forces that have been treated as real up to now, one would keep the terms which represent them, but one would regard these terms as expressing fictitious forces, inertial forces generated by hidden motions or constraint forces owed to the presence of hidden masses. In other words one would apply in a completely general manner the process employed by Maxwell for taking account of the electrodynamic actions. Thus one would obtain a Mechanics where there would once again be considered times, shapes, motions and masses, but from which the notion of force would have been rigorously removed; a Mechanics capable of satisfying atomistic philosophers and disciples of Gassendi and Huyghens.

"But", said Huyghens[1], "the greatest difficulty consists in showing how many different things are brought about by these principles alone."

It is, indeed, in following it through to the details of phenomena that one can appreciate exactly the value of a mechanical theory; such a doctrine, whose general principles are so beautiful and so logically inter-related, was lost in inextricable complications, in elusive subtleties when it wanted to compare the consequences of its deductions at least with natural laws. Newtonian Physics was an admirable edifice when Boscovich traced its whole plan; it collapsed when Poisson sought to construct from it the explanation of capillary phenomena.

Death did not leave Hertz the time to apply his general principle of Mechanics to particular problems. "He was obliged to assume", said Helmholtz[2], "that there existed a large number of masses not registering upon the senses, of invisible motions of these masses, in order to explain the existence of forces between non-contiguous bodies. Unfortunately, he gave no example capable of showing how he conceived of these intermediary terms. It is

---

[1] Huyghens: *Discours de la Cause de la Pesanteur*, (W. Burckhardt's Edition, Leipzig), Preface, p. 94.

[2] H. von Helmholtz: Preface to the work by Hertz: *Die Principien der Mechanik*.

evident that he would have been obliged to make an appeal to a considerable number of fictitious forces in order to give an account of the simplest physical actions."

This task, which Hertz was unable to accomplish has so far found no worker who has brought it to a successful conclusion. "In the consideration of the collision of molecules", said Boltzmann at the beginning of his *Lectures on the Theory of Gases*[1], "we shall preserve the old distinction between potential and kinetic energy. This distinction does not affect the nature of things. The assumptions which we shall make about the interaction of molecules during an impact have a completely provisional nature and will certainly give place to others later. I had at one time a temptation to outline a theory where the forces acting during the impact would be replaced by simple conditional equations (in the sense of the posthumous Mechanics of Hertz), more general than those of elastic impact; I gave it up because of arbitrary new assumptions that it would have also been necessary to make."

For want of having been applied to the solution of precise problems, for want of having been pursued to the point of determining the hidden masses and the hidden motions that would have to explain such and such a force taken wrongly as a real action, Hertz's Mechanics is, up to now, less of a doctrine than a project, more of a doctrine that a programme. This programme is itself reduced, in the last analysis, to this analysis: All the forces which are ordinarily introduced into the equations of Dynamics can be regarded as constraint forces owed to certain hypothetical bodies or as inertial forces produced by certain assumed motions. In order that this affirmation might have some scope, it would be good for it to be accompanied by an indication of a proper method for determining these bodies and motions when one knows the forces that they are called upon to replace. Now, this very indication is lacking.

Hertz's Mechanics therefore leaves completely undetermined the hidden motions and the hidden masses which must explain the forces of Nature. Under these conditions, how would one prove that a certain force is unexplained by these masses and motions? One could not find by experiment arguments that could convince Hertz's Mechanics of the error of the belief that it held.

---

[1] L. Boltzmann: *Leçons sur la Théorie des gaz*, (translated into French by A. Gallotti), (Paris), (1902), p. 3.

CHAPTER XIV

THE VORTEX ATOM

Hertz's Mechanics relieves the explication of the Physical World of the notion of force regarded as a primary and irreducible notion. Is this the term at which geometers necessarily have to be stopped in the long effort to reduce to a minimum the number of essential elements of every physical theory? Are they no longer able to drive their work of simplification further? The majority of the theories which have striven to explicate mechanically physical phenomena postulate the existence of small bodies that are indivisible and impenetrable, of atoms endowed with masses; could not this notion of atom endowed with mass, in its turn, lose its primary and irreducible character?

An answer to this question has been given by Kelvin; the progress brought to Hydrodynamics by Cauchy and Helmholtz had prepared this reply.

Let us consider a continuous medium in motion and, in this medium, a very small part of matter that our thought cuts out from the heart of that which surrounds it; at a given instant this particle offers a certain shape and occupies a certain position; at the end of a very short lapse of time it offers a shape a little different and occupies a position that is not quite the same. Cauchy has analysed the infinitesimal modification by which this material particle passes from the first state to the second; he decomposed this modification into elementary modifications each of which is very easy to conceive.

To lead a material particle from a certain state to another state very close to the first, one must first of all draw three certain mutually perpendicular lines, through one of the points that can be marked, which at the given instant are the *principle axes of dilatation* of the particle; upon the matter which forms it one imposes a first uniform and infinitesimal dilatation in the direction of the first axis, then a second dilatation in the direction of the second axis, and finally a third dilatation in the direction of the third axis; in general, these three *principal dilatations* are not all equal; their sum represents the *volumetric dilatation*; it is zero if the medium is incompressible.

The three principal dilatations successively impressed upon the particle impose upon it the change of shape that it must go through, it remains to analyse the alteration in position.

Through the point that has already been chosen one draws a certain line which for the moment considered is the *instantaneous rotational axis* of the particle, and one turns the entire particle around this line through a certain infinitesimal angle; dividing this infinitesimal angle by the infinitesimal duration of the total modification, one obtains the *instantaneous rotational velocity*.

Finally, one displaces the whole particle so that under this *translation* all its points describe infinitesimal courses, mutually equal and parallel.

Up to now we have only done Geometry, or, better, Kinematics; let us now come to propositions in Mechanics.

Let us imagine a continuous fluid, incompressible, non-viscous, of uniform and constant temperature; the motions of this fluid obey the equations that d'Alembert had deduced from his celebrated Principle and to which Euler gave a definitive form. Let us assume that the small parts into which our imagination can cut this fluid are subject to no force, or, at least, that they are not subject to those kinds of forces, brought into evidence by Clairaut, whose nature excludes every possibility of equilibrium for the fluid. In such a fluid the rotational velocities of the various particles obey some laws of a remarkable simplicity.

Here is the first, which was discovered by Lagrange[1]: If the instantaneous roational velocity of a certain particle is zero at an arbitrary instant of the motion, it always stays zero.

It was to prove this theorem completely rigorously that Cauchy[2] in 1815, formed some equations of extreme importance, but whose mechanical interpretation remained unseen for a long time; the use of a different method gave Helmholtz[3], in 1858, the key to this interpretation.

Being, at a given instant, continued from an extremely small quantity, the instantaneous rotational axis of a particle traverses a second particle contiguous to the second; the instantaneous axis of this second particle continued at the same instant by an

---

[1] Lagrange: *Mécanique analytique*, (2nd edn), Part II, Section XI, §1, art. 17.

[2] A. Cauchy: Mémoire sur la théorie de la propagation des ondes a la surface d'un fluide pesant de profondeur indéfinie, couronné par l'Academie des Sciences, *Mémoires des savants étrangers*, vol. I, (1827), p. 3; *Oeuvres de Cauchy*, vol. I.

[3] Helmholtz: Ueber Integrale de hydrodynamischen Gleichungen, welche der Wirbelbewebungen entsprechen, *Journal fur die reine und angewandte Mathematik*, vol. LV, (1858), p. 25; *Abhandlungen*, vol. I, p. 101.

infinitesimal distance is going to meet a third; and so on. We thus mark at this instant a row of particles, in the body of the fluid, contiguous to each other which follow the length of a curved line like pearls of a necklace along the thread that retains them; this curved line has as tangent at each of its points the instantaneous rotational axis of the particle this point belongs to; one then says that this curved line is, at the moment under consideration, a *vortex line*.

Now let us follow within the body of the fluid in motion the modifications of our row of particles; our necklace is deformed and displaced, it undulates in space without breaking the thread that links the pearls; and here is the essential property which is enjoyed by the curve drawn by this thread: just as it was a vortex line at the moment when it was traced, it remains a vortex line during the whole duration of the motion; the tangent that leads it from any one of its points always marks the instantaneous rotational axis of the particle that is at this point.

Let us take two particles a little separated from each other on a vortex line; let us observe them in their motion; at each instant let us measure, on the one hand the angular velocity of rotation which is perceptibly common to them, and on the other hand their mutual distance; these two quantities vary in the same sense; when the two particles accelerate their rotational motion they diverge; when they turn less quickly they approach; the ration of their instantaneous rotational velocity to their mutual distance remains constant.

Within the body of the fluid let us take a small surface at a given instant; through each point of the contour of this small surface let us consider a vortex line which passes through it at that instant; these lines are going to form the wall of a kind of very slender tube which extends into the fluid mass, now thickening, now shrinking; with Helmholtz let us give the name *vortex tube* to this kind of tube. The properties of vortex lines make us immediately perceive certain properties of vortex tubes and, in particular, the most essential one; indeed, it is clear that the fluid mass contained at a given instant in a vortex tube remains enclosed in a vortex tube indefinitely; the conservation of vortex lines implies the *conservation of vortex tubes*.

If, in a vortex tube, one takes two neighbouring cross-sections at a given instant, the material points comprising these two cross-sections will continually form two neighbouring cross-sections of the same vortex tube; after what we have seen, the mutual distance between these two sections will vary in such a way as to remain proportional to the angular velocity of rotation of the fluid that they contain between them; the volume they occupy, and which is that of the small cylinder lying between the two cross-sections, remains constant; the base of this cylinder therefore varies inversely as its height; thus we see that the part of a vortex tube which corresponds to a definite fluid mass swells when the angular velocity of rotation diminishes and deflates when the fluid turns more quickly; the area of the cross-section is inversely proportional to the instantaneous velocity of rotation.

This law assumes that one follows in time the same material portion of the vortex tube. One finds an analogous law upon inspecting the various parts of a vortex tube at the same instant; one established that this tube flares out in the regions of the fluid where the instantaneous rotational velocity is small and contracts in the regions where it is large; all along the same tube the product of the instantaneous velocity by the area of the cross-section keeps the same value.

This law implies a quite essential consequence: a vortex tube cannot end in the body of the fluid mass. In fact, in order that it could be constricted to the point that its cross-section were to become zero, it would be necessary for the angular velocity of rotation to become infinite at its end point. Hence it is necessary that a vortex tube traverses the whole fluid and culminates only at the very limits of this medium, or closes in on itself like a ring.

These remarkable theorems of Helmholtz led Kelvin, in 1867, to imagine the hypothesis of vortex atoms[1].

A unique form of matter fills the Universe; in its motions this homogeneous and incompressible matter obeys the laws that Euler's equations impose upon perfect fluids; in the beginning, some forces incompatible with the equilibrium of any fluid set this material in motion and created, in particular, a throng of vortex rings of all shapes and sizes; then when these forces disappeared, leaving in the world no more than apparent forces, explicable by the pressures and inertial forces of the universal fluid; these forces can neither generate a new vortex ring nor destroy one of those that pre-exist; each of these rings has become a true physical atom. The matter that impinges upon our senses is to be composed of such vortex rings.

This hypothesis of Kelvin presents us with the highest degree of simplification to which the explication of physical phenomena might be brought; not only is real force banished from the actual Universe, where we only establish apparent forces, effects of inertia and constraints, but yet the diversity that Chemistry believes to establish amongst simple bodies is only an illusion; it manifests to our senses only the different shapes and different motions taken by the vortex rings of a fluid everywhere identical to itself.

But Kelvin's hypothesis sank so deeply under ordinary perceptible things, that it became very laborious to get back to the latter and to furnish the explanation of facts that we establish every day. The simplest of them seemed to have no connection with the foundations of the theory. The fictitious forces that the fluid in the vortices generates do not account for universal gravitation; to explain it, Kelvin had to return to hypotheses similar to those of Lesage.

---

[1] W. Thomson: On Vortex Atoms, *Edinburgh Philosophical Society Proceedings*, (18th February, 1867).

The very Principles of Mechanics could not be deduced from the properties of vortex rings, and, as Maxwell remarked[1], one does not know where to discover in a vortex atom the constant element that would be suitable for considering its mass.

---

[1] Maxwell: Article: "Atom", *Encyclopaedia Britannica*; Brillouin: *Recherches récentes sur diverses questions d'Hydrodynamique; 1re partie: Tourbillons*, (Paris), (1891).

CHAPTER XV

GENERAL CONSIDERATIONS
ON MECHANICAL EXPLICATIONS

These difficulties, and plenty of others it would take too long to enumerate, tell us that it is time to stop; that it is not necessary to follow any further these attempts made to decrease more and more the number of primary notions upon which Physics rests. As well, the theory of vortex atoms has led us back quite close to Descartes' doctrines. The only body whose existence Kelvin admitted, this perfect, homogeneous, and incompressible fluid that filled the whole of space, that has no other property than that of moving in conformity with Euler's equations of hydrodynamics, is, to be sure, a close relative of this extension in length, breadth and depth, capable of all sorts of shapes and motions, which constitutes Cartesian matter.

One could press this assimilation still further, and certain people have dared to do so. Since Kelvin's fluid has no other property than that of supporting in space certain variable velocities according to certain formulae, why would not one go so far as to suppress it, as to denying it all substantial existence, as to reduce it to pure extension? The mass of a vortex atom, assuming that an acceptable definition could be found for it, would only be a symbol, only a mathematical expression combining the shape and motion of the atom; its constancy would not be the expression in mathematical language of the persistence of a material substance, but the consequence of a certain permanent distribution of rotational velocities; it is no longer true to say for this atom that "the physical law of the conservation of mass has degenerated[1] into a metaphysical axiom, the conservation of matter." From there, why should we not attribute more reality to the very matter of the fluid within the body of which the vortices are formed? Why should we not identify it with space, the con-

---

[1] W. Ostwald: La déroute de l'atomisme contemporain, *Revue générale des Sciences*, vol. VI, (1895), p. 954.

tainer of certain velocities and kinetic energies? Why should we not reduce Mechanics to the study of "extension and its alteration alone", alterations which leave unchanged the total quantity of Energy in the world? Would we not thus be led to the new doctrine, now in vogue, going under the name of *The Theory of the Migration of Energy*[1].

The moment we depart from the solid ground of traditional Mechanics to rush onto the wings of a dream in pursuit of this Physics which localises phenomena in an extension devoid of matter, we suffer an attack of vertigo; then with all our strength we cling to the bedrock of common sense; for *our most sublime scientific knowledge, in the final analysis, has no other foundation than the facts admitted by common sense*[2]; if one puts in doubt the certainties of common sense, the entire edifice of scientific truth totters upon its foundations and tumbles down.

We shall therefore persist with admitting that every motion assumes the existence of something moved, that every kinetic energy is the kinetic energy of some matter. "If you receive a blow from a stick", said Ostwald[3], "which do you resent, the stick or the energy?" We confess to resenting the energy of the stick, but we shall continue to conclude from it that there exists a stick, the bearer of that energy. Moreover, we shall not forget that this energy, which resides in certain places in space, which is transported from one region to another, singularly resembles Material, which might have changed its name but could not have changes its essence. And so we shall not stay on the side of doctrines for which the substantial existence of different and massive materials becomes an illusion, and we shall halt our discussions at the boundaries that Hertz himself did not cross.

The attempts made at explicating mechanically the physical phenomena the Universe presents to us fall nicely into two categories.

The attempts in the first category are carried out according to a method that can justly be named the *Synthetic Method*.

In this method one begins by constructing a mechanism from all the pieces; one says that some are the bodies which compose it, some are the shapes, the sizes, the masses, some the forces which act upon it; from these data one draws the laws according to which the mechanism moves, then by comparing these laws with the experimental laws that it is wished to explain, one judges whether there

---

[1] On the subject of this doctrine, see the preceding article by Ostwald, then: M. Brillouin: Pour la matière, *Revue générale des Sciences*, vol. VI, (1895), p. 1032; and: W. Ostwald: Lettre sur l'Énergétique, *Revue générale des Sciences*, vol. VI, (1895), p. 1069.

[2] P. Duhem: Quelques réflexions au sujet de la Physique expérimentale, *Revue des Questions scientifiques,* 2e série, vol. III, (1894).

[3] W. Ostwald: La déroute de l'atomisme contemporain, *Revue générale des Sciences*, vol. VI, (1895), p. 957.

is sufficient agreement between them.

For a long time this was the only method used. To it we owe the most celebrated examples of mechanical theories: the theory of magnetic attractions and repulsions given by Descartes; the explanation of weight by vortices, an essential doctrine of Cartesian Physics, which Huyghens perfected; the attempt of Fatio de Duilliers and Lesage to reduce gravitation to the impulse that material particles receive from other-world atoms; the theory of caloric, such as Laplace developed in his *Mécanique céleste*; the various kinetic theories of gases; Kelvin's gyrostatic aether; the cellular constructions by which Maxwell attempted to account for electromagnetic actions; the different mechanisms imagined in these latter years by Lorentz, Larmor, J.J. Thomson, Langevin, Jean Perrin, and yet others, to explain various effects of light, electricity and newly discovered radiations.

In every age since the renaissance of the Physical Sciences, but particularly in ours, this *Synthetic method* has run into the reluctance of certain minds; the adventurous character of the hypotheses upon which each of its explications rests; the somewhat puerile form of the mechanisms that it is obliged to imagine under the appearance of being sensible, have at all times presented a flank to plenty of sarcasm. "It is necessary to say on the whole: this is done by shape and motion", said Pascal. "But to say what is what, and construct the machine, that is ridiculous; for that is useless, uncertain and distressing." And Newton, launching his famous "*Hypotheses non fingo*", meant above all to throw the mechanisms of the Cartesians and Atomists out of his speculations.

In the eyes of the majority of physicists, the synthetic method no longer seems capable of giving a mechanical and complete explication of natural phenomena; it is, then, from the *Analytic Method* that one, today, requires such an explication.

The analytic method is what Maxwell defined in the preface of his *Treatise on Electricity and Magnetism* and that he endeavoured to put into practice in this *Treatise*. First of all he reduced to general formulae the laws of physical phenomena; then, without making any hypothesis about the nature of the motions by which these phenomena could be explained, he gave to these formulae an appearance which made their analogy with the equations of certain motions jump off the page.

If the formulae with which one has to do can be put into the form imposed by Lagrange upon the equations of Dynamics, things will go for the best. To the quantities which characterise the physical system subject to experimentation, one will be able to make correspond the variables and velocities which fix the shape and motion of a certain mechanical system, of such a king that the laws which preside over the transformations of the two systems are expressed by the same equations. The workings of the mechanical system will then explain the properties of the physical system.

If, furthermore, the formulae which condense the laws of experimentally studied phenomena let nothing fit the mould cast by Lagrange, the analytic method will not become ineffective, for all that; to assimilate these formulae into the equations of Dynamics it will assume that the system encloses some undetected masses and hidden motions; moreover, as nothing makes exact and limits the nature, number and complication of these masses and motions, it certainly seems that no kind of formulae could be taken as irreducible to the equations of dynamics; whatever these formulae may be, it is always permissible to hope that one could reduce them to the laws of Mechanics, either exactly or with such a degree of approximation as may be required.

There is more: the use of these masses and hidden motions will allow, if it be so desired, the suppression of every real force, the existence of only inertial and constraint forces; here again, the absolute indeterminacy left to the masses and hidden motions assures us that no geometer would stop our efforts at a solution of this problem by proving to us that this solution can be obtained neither exactly, nor approximately.

That the anayltic method therefore proposes simply to reduce the explication of physical phenomena to a Mechanics where the notions of motion, mass and force are taken as primary notions, or that it proposes to give this explication without appealing to the notion of force, does not mean to say that there is an experimentally found law from which one can prove that this method will obstinately refuse such an explanation.

Hence, *for physicists the hypothesis that all phenomena can be mechanically explained is neither true nor false; there is no meaning in saying so.*

Let us explain this seemingly paradoxical proposition.

In Physics one criterion alone allows the rejection as false of a judgement which does not imply a logical contradiction: the record of a flagrant disagreement between this judgement and the facts of experience. When a physicist affirms that this proposition has been compared with the experimental data; that amongst these data there were some whose agreement with the proposition being tested that were not necessarily *a priori*; that, however, the discrepancies between these data and this proposition remained below certain experimental errors.

In view of these principles one does not state a proposition that Physics takes to be erroneous by advancing the idea that all the phenomena in the inorganic world can be explained mechanically; for experiment yields to us no phenomenon that is assuredly irreducible to the Laws of Mechanics. But it is no longer legitimate to say that this proposition is physically true; for the impossibility of forging back to a formal and insoluble contradiction with the results of observation is a logical consequence of the absolute indeterminacy that is left to the invisible masses and hidden motions.

Thus it is impossible for anyone who holds to the processes of the experimental method to declare as true this proposition: *All physical phenomena are explained mechanically*. It is also impossible to declare it false. *This proposition transcends the physical method.*

If, in regard to this proposition, one wishes to depart from a state of mind where every decision remains suspended, one will have to resort to arguments unknown to experimental method.

These arguments could be of two kinds; they could consist of arguments drawn from Metaphysics; they could also invoke convenience as a preferential ground, repudiating every pretence at philosophy.

It is by metaphysical arguments that Descartes established the necessary reduction of all physical phenomena by some 'mechanical arguments'; it is because there is no clear idea in the notion of substance, if it is not those that geometers have been accustomed to see, that it makes extension into length, breadth and depth the very essence of matter; it is because matter is essentially identical with the space treated by geometers that one has nothing to embody in a wholesome Physics if it is not various shapes and motions; it is evident that it is the same thing to raise one book through two hundred feet or two books through one hundred feet, and it is upon this evidence that the whole of Statics if founded; the divine immutability assures us that the Creator always keeps in His work the same quantity of motion that He put into it at the beginning, and this conservation of the quantity of motion is the first Principle of Dynamics.

Descartes' Dynamics, drawn from metaphysical arguments, scarcely agreed with Galileo's discoveries about falling heavenly bodies; and soon Leibniz, substituting the conservation of kinetic energy for the conservation of momentum, entitled his argument: *Demonstratio erroris memorabilis Cartesii*. Since the refutation of this memorable error I think no philosopher worthy of the name has attempted to draw the primary principles of Mechanics from Metaphysics; it is clear to all that it is experiment alone that, by its testing, guarantees the value of these principles; Metaphysics, which is recognised as being incapable of justifying them, could say nothing about whether their empire is restricted to perceptible motions alone or if it extends to the set of physical phenomena.

Thus the metaphysical method, no more than the physical method, is unable to anser the question: Is it true or false that all physical phenomena are reducible to local motions subject to the Laws of Dynamics?

Any number of us will renounce the question so formulated, having no answer, and substitute for it this other question: Is it helpful to one who wishes to propound Physics, is it useful to one who wishes to increase its coverage, to reduce all physical phenomena to motions, to reduce all physical laws to the equations of Mechanics?

In this new form the question loses the absolute character it has had to date; it is now clear that different physicists will be able to give different answers to it, without logic itself having to reduce any of them to silence.

GENERAL CONSIDERATIONS

The degree of suitability of a method, in fact, is essentially a matter of personal appreciation; the particular turn of each thinker, the education received, the traditions immersed in, the customs of the environment in which he lives, all influence this appreciation to a high degree; these influences vary in the extreme from one physicist to another; also, could not one consider an exposition of Physics to be infinitely elegant and easy which the other would judge to be heavy and highly inconvenient.

When one examines the attitude of various thinkers to physical theories, one can classify them in two large categories: the category of *abstract* ones and the category of *imaginative* ones.

The *abstract thinkers* were content to consider clearly defined quantities, furnished by defined measuring processes able to enter into rigorous arguments and precise calculations according to fixed rules; it mattered to them little that these quantities cannot be imagined. They are satisfied, for example, if they have defined a thermometer which to every intensity of heat brings into correspondence a determined degree of temperature, or if they know the form of the equations which linked this temperature to the other measurable quantities of substances, density, pressure, latent heat of fusion, the latent heat of boiling. They do not require this temperature to reduce to the kinetic energy of an imaginable motion enlivening the molecules whose behaviour it is meant to depict. Provided that the laws of Physics may be condensed to a certain number of abstract judgements expressible as mathematical formulae, they willingly agree to these judgements bearing certain ideas foreign to Geometry. They are resigned to the physical world not being susceptible to a mechanical explication, with no trouble.

The *imaginatives* have quite other requirements. For them, "the human spirit[1] observing natural phenomena, recognises in them, besides plenty of confusing elements that have not succeeded in being unravelled, one clear element that, by its precision, is able to be the object of truly scientific attainments. This is the geometric element pertaining to the localisation of objects in space, and which allows them to be represented, depicted or constructed in a more or less ideal way. It is constituted of the dimensions and forms of the bodies or of the systems of bodies, which is what one calls, in a word, their *configuration* at a given moment. These forms, these configurations, whose measurable parts are distances or angles, whilst being conserved, at least nearly so, during a certain time and even seen to be kept in the same regions of space to constitute what one calles the *resting place*, whilst changing incessantly, but continuously, and their changes of position are what are called the *local motion*, or simply the motion."

These different configurations of bodies, their changes from one instant to another, are the only elements that a geometer can

---

[1] J. Boussinesq: *Leçons synthétique de Mécanique générale*, (Paris), (1889), p. 1.

portray; these are also the only ones that the imaginative thinker can represent clearly; according to him, they are therefore the only ones that must be the objects of science. A physical theory will only be constituted when it has led from the study of a group of phenomena to the description of such and such figures and such and such local motions. "Up to now, science[1], considered in its turn as built, or susceptible of being so, has grown going from Aristotle to Descartes and to Newton, from the ideas of *qualities* or *alterations* of state, which are not described, to the idea of *forms* or local *motions* which are described or are seen."

An imaginative physicist will therefore not regard anything as satisfactory as long as it has not replaced the various qualities of bodies, understandable only in the abstract and by arithmetic representation, by combinations of figure comprehensible to geometric intuition and capable of being drawn.

Are the theories which have been proposed up to now for explicating physical phenomena mechanically going to provide for his imagination the representations delineated beyond which, for him, there is no clarity.

Yes, assuredly, it is a question of the ancient mechanical theories formed by the synthetic method. At the very root of such a theory there are definite hypotheses about the shape of atoms and molecules, about their size, their arrangement; it is sufficient to open a book where there is such an explication, that this book contains the name of Descartes or Maxwell, to find in it some drawings showing the appearance that the texture of substances offer after sufficient penetration.

But the explicatory value of the mechanical theories formed by synthesis today seems quite dubious. Too clearly does it appear that each of them is appropriate, at the most, for representing a miniscule fragment of Physics; that these fragmented representations may not be welded to each other to form a coherent and logical explication of the inanimate Universe. One then has recourse to the analytic method; one groups into a set of mathematical formulae the laws which the corporeal qualities and their alterations obey, and one attempts to prove that this set of formulae is not incompatible with a mechanical explication of physical phenomena.

This process — does it render such an account? — no longer provides any food for the imagination, eager to promote reason, even to exceed it, in the understanding of physical phenomena; it no longer satisfies the desires of him who, in qualities and their alterations, wants to lay hold of something which may be drawn or seen.

In the first place, this analytic method assures that although the physical laws established are not incompatible with a mechanical explication, it does not lead us to know the detail of this

---

[1] J. Boussinesq: *Théorie analytique de la Chaleur*, vol. I, (1901), p. xv.

GENERAL CONSIDERATIONS

explication in an explicit way; it asserts "on the whole that this is done by shape and motion", but it does not tell us by what shapes nor by what motions, it does not "construct the machine"; it does not even indicate how it could be constructed; it gives no process for extracting from the equations it studies the plane of a mechanism capable of working in accordance with these equations. How could the masses and motions which stay *hidden*[1] be better received by imaginative thinkers than the *occult*[1] powers of the ancient Scholastics.

In the second place, the analytic method brings into evidence this truth: If one can construct a machine capable of explaining a set of physical laws, one can construct an infinity of others which will explain just as exactly the same set of laws. "Hence, if a phenomenon has a complete mechanical explication[2], it will have an infinity of others which give an equally good account of all the particularities revealed by experiment." Amongst all these mutually equivalent explications, equally acceptable to an abstract thinker, the imaginative physicist's mind will float, hesitating, looking to see how to decide upon a convincing argument that could never be discovered, and finding only motives that have nothing general nor absolute to guide his choice.

Finally, if the analytic method ensures that the set of physical phenomena is capable of a mechanical explication, it also and particularly allows one to foresee that this explication, to be complete, will have to invoke a prodigious multitude of invisible masses, an infinite complexity of hiddent motions; and one would divine that the most powerful imagination, far from picturing accurately the world's mechanism, would be driven to distraction in a seeming chaos.

Hence the analytic method, which alone seems capable of providing from the Laws of Physics a logically constructed mechanical explication, seems incapable of satisfying the requirements of imaginative physicists, that is to say, of the very ones who required a mechanical interpretation of phenomena.

If these physicists want, at any price, to picture the qualities of bodies in shapes suitable for geometric intuition, in shapes simple enough to be depicted in a table clearly understandable to the eyes and the imagination, they will have to renounce the hope of uniting all these representations into a coherent system, into a logically ordered science. It will be necessary "for each to choose[3] a way of reasoning about the world which is a correct as possible — and which particularly is *rapid, intuitive* and *fertile*."

Many gave it up. They renounced classifying the different natural laws actually known into a sequence all of whose terms are linked to each other with an irreproachable order and perfect

---

[1] Hidden ... hidden: *Fr: cachés ... occultes.*
[2] H. Poincaré: *Électricité et Optique,* vol. I, (Paris), (1890), Introduction, p. xiv.
[3] M. Brillouin: Pour la matière, *Revue générale des Sciences,* vol. VI, (1895), p. 1034.

rigour; they prefer to simulate mechanisms whose operation simulates more or less exactly phenomena already discovered and, sometimes, help new ones to be suspected. They then returned to the synthetic method but without demanding from it the one well-coordinated Physics which it cannot provide. To each category of phenomena they make correspond an arrangement of shapes and motions which are a more or less happy imitation, in the words of English physicists[1], the *model*. This model they construct from items as concrete, as accessible to our senses and imagination as possible; Kelvin did not hesitate to bring into his schematic constructions *strings* and *reverberations of small bells*; it is no longer a question, in fact, of conceiving of a mechanism which could be regarded as the expression of reality, as the reflection of the *quid proprium* of material things; to a mind that does not grasp pure abstraction, it is a question of obtaining help from objects which may be touched and seen, which may be carved and drawn.

Not only must the elements which compose a model be easy to imagine and, for that, resemble as much as possible the bodies we see and handle every day, but, further, these elements must be less numerous; the arrangements in which they are combined must be relatively simple. This simplicity, wanting which it would cease to be useful, forbids to the model the pretention of representing a far-reaching set of natural laws; the use of a definite model is necessarily very restricted; each chapter of Physics required the construction of a new mechanism, unconnected with the mechanism which served to *illustrate* the preceding chapter.

Reduced to illustrating each group of phenomena by models, Mechanical Physics is able to remain a precious aid, for certain thinkers, without which the laws, when formulated as abstract propositions, would be less easily and less fully accessible to them; it can excite the curiosity of several, and, by way of analogy, suggest some discoveries to them — such as Lorentz's electro-optic model which led Zeemann to recognise the action of a magnetic field upon the lines of a spectrum. The use of models can even become indispensible to certain geometers whose faculty of abstraction is not so powerful as their imagination; and amongst the latter one can count some of the greatest physicists of the age, who subscribe to Kelvin's words: "It seems to me[2] that the true meaning of the question 'What do we understand or not understand by a particular subject in Physics?' is this: Can we make a corresponding mechanical model? ... I am never satisfied[3] as long as I have been unable to make a mechanical model for the object; if I can make a mechanical model, I understand; for as long as I cannot make a mechanical model, I do not understand."

---

[1] Touching the constant use that the English make of the *model* to *illustrate* physical theories, see the article on *l'École anglaise et les théories physiques* which we have published in the *Revue des Questions scientifiques, 2e série*, vol. II, (1893).

[2] W. Thomson: *Lectures on Molecular Dynamics*, p. 132.

[3] W. Thomson: *ibid.*, p. 270.

GENERAL CONSIDERATIONS

Such intellectual requirements, such an identification of the two verbs *to understand* and *to imagine*, greatly surprise — I would almost dare to say scandalise — those who can conceive of an abstract idea without the help of geometric or mechanical representations; the latter, however, must not deprive of this help those whose minds' nature calls out for it; they can only repeat Helmholtz's wise words[1]: "English physicists, for example, Lord Kelvin, in his theory of vortex atoms, and Maxwell, in his hypothesis of systems of cells with rotating contents, on which he bases his attempt at a mechanical explication of electromagnetic processes — have evidently derived a fuller satisfaction from such explanations than from the simple representation of physical facts and laws in the most general form, as given in systems of differential equations. For my own part, I must admit that I have adhered to the latter mode of representation and have felt safer in so doing; yet I have no essential objections to raise against a method which has been adopted by three physicists of such eminence."

These concessions reach, if not exceed, the furthest limit that can be accorded to the use of mechanical models in Physics. The legitimacy of this use is of a purely practical and not a logical order. A sequence of disparate models cannot be regarded as a physical theory, for it lacks in it the very essence of a theory, the unity that links the laws of the different groups of phenomena into a rigorous order. Could it not be given *a fortiori* as an *explication* of the facts that are observed in the inorganic world; it can offer nice analogies, intuitive, fertile, between the Laws of Physics and the behaviour of certain mechanisms; yet, as an old adage has it, *comparison is not proof*[2].

And so those who are resigned to the use of mechanical models note carefully that they renounce "conceiving of the cause of all natural objects by arguments in Mechanics"; either they regard such an explanation as too complicated to be manageable and fertile, or they have even ceased to believe it to be possible.

---

[1] H. von Helmholtz: Preface to the work of H. Hertz: *Die Principien der Mechanik*, p. xxi.
[2] *Fr: comparaison n'est pas raison.*

PART TWO

THERMODYNAMICAL THEORIES

CHAPTER I

THE PHYSICS OF QUALITY

To attempt to reduce all the properties of substances to shape and motion seems a chimerical enterprise, either because such a reduction would be obtained at the price of complications that would scare our imagination away, or even because it would be in contradiction with the nature of material things.

And so we are now obliged to accept into our Physics something other than the purely quantitative elements treated by geometers, to admit that matter has *qualities*; at the risk of us being understood to begrudge the return to *hidden properties*, we are compelled to keep as a primary and irreducible quality those by which bodies are hot, lighted, electrified or magnetised; in a word, renouncing the attempts that have been repeatedly begun since Descartes, we have to refasten our theories to the most essential notions of Peripatetic Physics.

Will not this return to the past compromise all the body of doctrine raised up by physicists after they had shaken off the yoke of the Ancient School? Will not the most productive methods of modern Science fall into disuse?

Convinced that all in corporeal Nature is reduced to shape and motion such as geometers conceive, that all is purely quantitative, physicists introduced measure and number everywhere; every property of bodies became a quantity; every law an algebraic formula; every theory a linking of theorems. Admirable in its precision and rigour, of a majestic unity, Physics was then the 'Universal Mathematics' dreamt of by Descartes. Would it be necessary to break this perfect form, at one and the same time so convenient and so beautiful? Would we have to rebuff the marvelously powerful aid that the use of numerical symbols provided for our deductions? Would we become resigned to the vague discourses and confused shadowy quarrels which constituted Natural Science before academics made use of algebraic language. Shall we have to face once again the sarcasm which discredited the Ancient School's

Cosmology? With one accord no physicist would consent to it.

No such sacrifice is necessary. The abandonment of mechanical explications does not have the consequence of abandoning Mathematical Physics.

As is already known, numbers can serve for representing the various states of a quantity that is additive; the step from quantity to number which it represents properly constitutes the *measurement*. But the number can also serve to locate the various intensities of quality. This extension of the notion of measurement, this use of the number as the symbol for a thing that is not quantitative, undoubtedly astonished and scandalised the Peripatetics of Antiquity. In that lies the most certain progress, the most durable conquest which we owe to the seventeenth century physicists and those who continued; in their attempt everywhere to substitute quantity for quality they ran aground; but their efforts were not in vain, for they had established the following priceless truth: *It is possible to discourse about physical qualities in the language of algebra.*

An example will show us how this passage from quality to number is carried out.

The sensation of heat which we experience upon touching the various parts of a substance enables us to perceive a quality of this substance; it is what we express by saying that this substance is hot. Two different substances can be equally hot; they possess the quality considered with the same intensity. One of two bodies may also be hotter than the other, and then the first possesses the quality considered with more intensity than the second.

Without going more deeply into the nature of the quality expressed by the adjective *hot*, above all without resolving it into quantitative elements, we can very well conceive that a number is caused to correspond to each of its states, to each of its intensities; that two equally hot bodies are characterised by the same number; that of two unequally hot bodies the hotter is characterised by the greater number; the numbers thus chosen will be *degrees of temperature*.

These simple indications already show us how, in place of discussing *hot* in ordinary language, one will be able to apply the symbols of algebra to *degrees of temperature*; instead of saying that a body is as hot as, hotter than or less hot than another, one will write that the degree of temperature of the former is equal to, greater or less than the degree of temperature of the latter.

From now on it will be understood that a theory where *hot* will be presented, will no longer be done in the form of a philosophical exposition in the manner of those scholastic dissertations where confusion and obscurity slip in so easily, but in the form of a sequence of equations or algebraic inequalities, offering the highest degree of clarity and precision that the human mind can attain.

It is not sufficient that the use of algebraic signs allows us to treat hot with clarity and precision, but abstractly and generally; it is still necessary that we should assure the passage of our abstract and general propositions to concrete and particular verities, that we may compare the consequences of our theories with experimental data; for testing against facts constitutes, for a physical theory, the unique criterion of truth.

This passage from the abstract to the concrete, from the general to the particular would be impossible if we knew only that at each intensity of heat of a body one could make correspond one degree of temperature, and that this degree rose when this intensity increased. It would further be necessary that a practical rule provided us with the numerical value of the degree of temperature of an exactly specified body, that a certain *instrument*, put into relation in a specified way with the body whose degree of temperature we wish to know, recorded this degree for us. The mathematical formulae in which the letter T, the symbol of temperature, appears, only take a physical meaning through the choice of a *thermometer*.

The employment of the thermometer chosen is subject to certain rules, subject to certain conditions; it requires, for example, that the temperature of the body under experimentation be uniform, that it remain constant during a certain time, that it be neither too high nor too low. The indication of a thermometer that may be as perfect as one likes, is not exact, but approximate; to two different, but too close, intensities of heat, this instrument does not make correspond two discernible indications; to a given intensity of heat it does not bring into correspondence a unique degree of temperature, but all the degrees of temperature that lie between two certain limits whose separation eludes our means of observation.

Hence one will not be able to compare with experiment, by means of a thermometer, all the consequences of the theory, but only certain of them; thus, those which have the character of temperatures varying from one point to another or from one instant to another, or those that concern too hot or too cold bodies will remain outside direct testing. In the very case where comparison will be possible, it will not be absolutely rigorous; its exactness will be limited and will depend upon the degree of precision of the thermometer. Nevertheless, this instrument will permit the passage from abstract and general propositions that the theory formulates, to the concrete and particular judgements that experiment furnishes; this passage will be possible in cases all the more extensive as one makes all the wider the conditions under which the use of the thermometer is legitimate; this transition will be made with all the more sureness as the thermometer becomes all the more precise. By the definition of and the use of an instrument, the theory assumes a physical meaning; it becomes verifiable and usable.

What we have just said touching upon the quality that consists in *being hot*, and touching upon its symbolical representation by a number, the *degree of temperature*, can be repeated, *mutatis mutandis*, for all the qualities which attract physicist's attention: electrification, magnetisation, dielectric polarisation, illumination[1,2]. Analysis of the experimental facts leads us to conceive of the abstract notion of quality that is the more or the less intense; to this quality we bring into correspondence a numberical symbola whose value is thelarger the more intense the quality is; this correspondence, whose possibility is asserted quite generally, is practically assured in extended cases by the use of an instrument; this instrument determines approximately the numerical value of the symbol which corresponds to a quality specified in reality. Without a *measurement process* the definition of the physical quantity which symbolises a quality would be incomplete and deprived of meaning; only does this process assure the passage from the general and abstract algebraic formula, by which a Physical Theory is expressed, to qualitative fact, particular and concrete, to which one wishes to apply this law.

Half a century ago Rankine[3] had already sketched these principles in some pages that are too little known; they laid bare the very structure of this strange science that is Physics, *the experimental science of corporeal qualities*, and, yet, *a science which is developed in a sequence of algebraic calculations*.

The geometers of the scientific Renaissance did not reproach only the Physics of the Ancient School for its lack of precision, which had avoided the use of algebraic language; they reproached it also, and above all, for creating so many hidden properties, substantial forms, sympathies and antipathies that are met in the world of effects to be explained; and so they accused them of degenerating into a verbiage whose bloated form stimulated the vanity of pedants and the admiration of dolts, but whose basis, sunken and empty, provided no nourishment for the curiosity of exact and considered thinkers. It is not necessary for the new Physics to merit this reproach.

Physics will therefore reduce the theory of phenomena presented by inanimate Nature to the consideration of a certain number of qualities; but it will seek to reduce this number to the smallest possible one. Each time a new effect is presented it will attempt to reduce it to the qualities already defined in all ways; it is only after having recognised the impossibility of this reduction that it will be resigned to admitting a new quality into its theories, to introducing a new kind of variable into its

---

[1] *Fr: éclairement*.

[2] On the subject of the representation of the quality that is meant by the words 'to be lightened' (*Fr: être éclairé*) by means of mathematical symbols proper to the building of a theory of light, we refer the reader to our: Fragments d'un cours d'Optique, Ann. de la Soc. Scient. de Bruxelles, vols. XVIII, XIX, XX, (1894-1896).

[3] J. Macquorn Rankine: Outlines of the Science of Energetics, *Glasgow Philosophical Society, Proceedings*, vol. III, No. 6, (2nd May, 1855), *Miscellaneous Scientific Papers*, p. 209.

equations. Thus the chemist discovers a new substance and endeavours to decompose it into some elements already known; it is only when he has exhausted, in vain, all the means of analysis available to laboratories, that he decides to add a name to the list of elementary substances.

The name *elementary* is not given to a chemical substance by virtue of a metaphysical argument proving that it is indecomposable in nature; it is given to it by virtue of a fact—because it has resisted all attempts at decomposing it. This adjective is an admission of powerlessness; a substance, elementary today, will cease to be so tomorrow if some chemist, more fortunate than his predecessors, manages to dissociate it; potash and soda, substances elementary for Lavoisier, became compounds after the work of Davy. Thus it is for the primary qualities that we admit in Physics. In calling them *primary* we do not prejudge them as irreducible in nature; we simply avow that we do not know now to reduce them to simpler qualities; but this reduction, which we cannot do today, will perhaps tomorrow be a *fait accompli*. Illumination, for example, appears in the beginning of Optics as a primary quality; one day, perhaps near at hand, when the electromagnetic theory of light will triumph definitively, illumination will be reduced to rapid changes in another quality, of dielectric polarisation; it will then lose its rank as a primary quality.

The number of primary qualities received into Physics must be as small as is consistent with our actual knowledge, just as the number of elementary substances admitted in Chemistry is the smallest possible, given our means of analysis. This number does not exceed eighty by much and it increases incessantly with the discovery of new elements. So one must not be surprised if the list of primary qualities is long and if the unceasing labours of physicists decorate history from time to time with the contribution of a new quality.

The theories of Mechanical Physics set themselves up as explications of the material world; using the appearances and qualities revealed to us by experiment, they aspired to dissecting the intimate structure of substances and uncovering the ultimate reasoning for their properties. It is certain that the New Physics would not have similar pretensions. When it adds a certain property to the number of primary qualities, it acts from modesty; it does not pretent to an explication, it confesses its powerlessness to explain. By substituting a numerical symbol for a quality revealed by experiment, it does not add a new lesson to the teaching of experiment; similarly, by expressing an idea, language does not enrich the content of this idea; the calculations to which the degree of temperature can be submitted will teach us nothing, touching upon the intimate nature of the quality represented by this degree, that a close study of our senses or of observational data does not teach us. The New Mathematical Physics does not pride itself in penetrating, in our knowledge of corporeal qualities, beneath what the analysis of experimental facts reveals to us; briefly, it is a *Physics*; it is not a *Natural Philosophy*, a *Cosmology*, a branch of *Metaphysics*.

If the Theoretical Physics renounces giving an explication of the material world, what will be its rôle and its object, therefore? The formulae that it substitutes for experimental laws will in each particular case allow one to replace the letters which appear in such a formula by the numerical values which fit the properties of the concrete substances studied; having carried out this substitution, the application of the general law to the particular case will be done with a rigour and thoroughness that is limited only by the degree of exactness of the instruments; finally, these formulae will be condensed into a small number of very general principles, from which they will be able to be deduced by Analysis and algebraic calculations; the logical order into which our knowledge of Physics is then classified will make a system that is easy and sure in its use; it will allow a physicist rapidly to find, without error or omission, all laws upon which the solution of a given problem depends.

Our senses perceive only the surface of reality; this surface covers depths which will, doubtless, always stay unknown to us; which, probably, we would be unable to understand even if some higher intelligence wished to reveal it to us, nor to express it if, having once understood it, we wanted to make it known to our fellow men; which, furthermore, would perhaps be unusable if we were to conceive of it, for our means of action, being coordinated to our means of cognition, would no more modify the essence of substances than the understanding of them. The New Physics will no longer have the object of discovering to us this foundation of things; its aim will be more modest, and more practical at the same time. This aim will be to assist our activity of taking possession of the material would, to modify it, to enslave it to our needs; it will consist of rendering more robust or delicate the tools by means of which we can mould substances, to diversify these tools in order that each of them be better adapted to its object, and finally to classify them methodically, so that the physicist may lay hold, without having to grope and without delay, upon whatever suits his task.

CHAPTER II

ON THE COMPARISON OF THEORY
WITH EXPERIMENT
AND ON VIRTUAL MODIFICATION[1]

Three distinct domains are simultaneously present for the physicist's consideration.

The first is the *domain of experimental facts*; these facts, produced in the external world, are recorded by a physicist's senses, and his faculty of generalisation and induction recognises the laws from them.

The second is the *domain of theory*; it is a set of quantities and symbols whose algebraic properties have been defined and which are engaged in a system of properties and formulae logically deduced from a small number of fundamental postulates.

The theoretical domain has the object of offering a symbolic description, a *scheme* as extended, as complete and as detailed as possible, of the domain of experimental facts. So that the theory may not be a language stripped of meaning, pure play with formulae, it is necessary that there be a key to make symbols correspond with reality, the indicator of the thing signified,

---

[1] *Translator's Note:* Duhem's word 'modification' is used here in preference to the more commonly used term 'change'. The reason for this preference is straightforward: It will be seen that 'modification is primarily used in the context of infinitesimal alterations of a condition to yield another condition that differs from the former almost insignificantly. Such infinitesimal modifications are also combined to produce finite alterations by a smooth process. A 'modification' may also be a discrete alteration caused by a non-reversible process. However, the term 'change' does not so readily imply, especially tacitly, all the connotations of infinitesimal alterations, their combination to form smooth processes of change, and even irreversible change, that the word 'modification' so happily succeeds in doing. Add to this explication the note that Duhem almost invariably used the term 'modification' throughout his writings.

it is necessary to be able to translate theoretical formulae into experimental facts. The study of this key belongs to the third domain imposed upon the physicist, to the *domain of instruments and measuring processes.*

As for the inter-relation of these three domains, how many important remarks there are to be made![1] We can only indicate a small number of them, selecting those that are essential for the comprehension of the New Mechanics.

These remarks concern the laws that preside over the development of an exact theory.

The materials with which this theory is constructed are, on the one hand, the mathematical symbols that serve it for representing the various quantities and the various qualities of the physical world; on the other hand, the general postulates which serve it as principles. With these materials it must be raised into a logical edifice; it is therefore required to respect scrupulously the laws that Logic imposes upon all deductive reasoning, the rules that Algebra prescribes for every mathematical operation.

The mathematical symbols of which the theory makes use have a meaning only under well defined conditions; to define these symbols is to enumerate these conditions. Other than under these conditions, will the theory never use these signs. Thus by definition, an absolute temperature can only be positive, the mass of a body is invariable; never, in these formulae will absolute temperature be given a zero or negative value; never, in its calculations, will the mass of a given body be varied.

For its principles, the theory has some *postulates*, that is to say some propositions that may be stated just as one may please, provided that there is neither a contradiction within the terms of one and the same proposition, nor between two distinct propositions. But, once these postulates have been posed, it is held that they should be kept to with a rigorous jealousy. If, for example, it took the principle of the conservation of energy as the basis of its arguments, it must forbid every affirmation out of tune with this principle.

These rules are imposed with all their weight upon a physical theory that may be constructed; one single lapse would render it absurd and would constrain us to reject it; they alone are but imposed. IN THE COURSE OF ITS DEVELOPMENT, *a physical theory is free to choose the path it pleases, provided that it eschews all logical contradiction; in particular, it has to take no account of experimental facts.*

This is no longer so WHEN THE THEORY HAS ATTAINED ITS COMPLETE DEVELOPMENT. When the edifice has reached its peak, it becomes necessary to compare with the set of experimental facts the set of propositions obtained as conclusions from these long deductions; it is necessary to be assured, by means of employing some adopted measuring processes, that the first set finds in the second an

---

[1] P. Duhem: Quelques réflexions au sujet de la Physique expérimentale, *Revue des Questions scientifiques,* 2e série, vol. III, (1894).

image sufficiently resemblable, a symbol sufficiently precise and complete. If this accord between the conclusions of the theory and the experimental facts were not manifested to a satisfactory approximation, the theory would have been able to have been constructed logically; it would nonetheless have to be rejected because it was contradicted by observation, because it would be *physically* false.

This comparison between the conclusions of the theory and the facts of experience is therefore indispensible, since alone the control afforded by observation can give to the theory a physical value; but this control, exclusively, can strike at the conclusions of the theory, for alone do they offer an image of reality; the postulates that serve as a point of departure for the theory, the intermediaries through which one passes from postulates to conclusions have not been submitted to it.

When, therefore, in the course of the deductions from which the theory flows, one submits to some algebraic operations and to calculations the quantities with which the theory deals, one has not wondered whether these operations, if these calculations *have a physical meaning*; to speak more explicitly, one has not wondered whether the employ of measuring processes would allow them to be translated into concrete language, and whether, translated thus, they would correspond to real or possible facts. To pose a similar question would be to conceive of a completely erroneous notion of the structure of a physical theory.

Here do we touch upon a principle so essential, and, at the same time, so subtly perceived, that it will be permitted to us to insist upon explaining our thought by an example.

J. Willard Gibbs has studied theoretically the dissociation of a perfect gaseous composition into its elements, each regarded as a perfect gas. A formula has been obtained that expresses the law of chemical equilibrium within the body of such a system. I propose to discuss this formula. To this end, leaving unchanged the pressure that supports the gaseous mixture, I consider the absolute temperature that appears in the formula, and I make it vary from 0 to $+\infty$.

If to this mathematical operation one wishes to attribute a physical meaning one is presented with a veritable mob of objections and difficulties. No thermometer can be made to recognise temperatures below a certain limit, none can determine sufficiently high temperatures; this symbol which we have called *absolute temperature* cannot, by the measuring processes at our disposal, be translated into something that has a concrete meaning, unless its numerical value stays between a certain minimum and a certain maximum only. Furthermore, at sufficiently low temperatures this object that Thermodynamics calls a *perfect gas* is no longer the image, even an approximate one, of any real gas.

These difficulties, and plenty of others that it would take too long to enumerate, vanish when one takes note of the remarks that we have formulated. In the construction of a theory the discussion of which we have just spoken is only an intermediary;

it is not correct to seek a physical meaning for it. It is only when this discussion will have led us to a series of propositions that we shall have to submit these propositions to the inspection of facts; then we shall examine whether, between limits where absolute temperature can be translated into concrete thermometric indications, where the idea of perfect gas has been closely realised by the fluids that we observe, the conclusions of our discussions agree with the results of experiments.

These principles bring to the full light of day a notion that will play an essential role in all the development of Theoretical Physics, the notion of *virtual modification*.

In the mathematical schema under which Theoretical Physics proposes to depict reality, the material system that is required to be studied is represented by a whole procession of mathematical quantities which measure its various quantitative elements or those which give reference to its various qualities. Amongst those quantities there are some that their very definition renders incapable of variation; thus the mass of a specified body, the charge on an isolated conductor, may not vary. Others, on the contrary, are able to change their value. Such variations are subject to no restrictions that arise from their definition; thus without contradicting the definition of the intensity of magnetisation at a point within a magnetic medium, one can attribute to this intensity every magnitude and every direction. There is also that of which the capacity to vary is restricted by certain *constraint conditions* which flow from their very definition. These conditions can be inequalities: within the body of a mass of water susceptible of freezing, but that contains no fragment of ice, the mass of ice can increase, but it cannot diminish. These conditions can also be equalities: in a system that contains calcium carbonate, lime and carbon gas, there is a ratio between the mass of lime and the mass of carbon gas that can appear simultaneously or disappear simultaneously.

To impress upon the variable quantities that characteristic state of a system some infinitesimal alterations allowed by the constraints is to impose upon the material system a virtual modification.

It is thus to produce a virtual modification that the position of mobile bodies is infinitesimally altered, the shape of deformable bodies; but it is also to produce a virtual modification that one lowers or raises infinitesimally the temperature, to change by an infinitesimal proposition the magnitude and direction of the magnetisation at each point of a mass of iron, to modify infinitesimally the electric distribution on a conducting body, to fuse an elementary mass of ice, to freeze, to vapourise an elementary mass of water, to submit a composition to an infinitesimal dissociation, to produce the combination of infinitesimal quantities of two bodies.

The use of these virtual modifications is an artifice of reasoning, a calculational process; it is therefore useless for a virtual modification to have a physical meaning: I shall take, at a given pressure and a given temperature, a mixture of oxygen, of hydrogen and of water vapour. By these same words I can understand two things quite distinct: I can understand, in the first place, a concrete mixture of three real fluids, enclosed in a certain receptacle of glass or of porcelain, in relation with a manometer outside the laboratory cabinets heated by bunsen burners or by a reverberation furnace. I can understand, in the second case, a schematic system of symbols and quantities, the description of the concrete system; in this schematic system, the oxygen, the hydrogen, the water vapour, are no longer colourless, odourless fluids contained within a receptacle, but groups of letters, O, H, H_2O, accompanied by a procession of numbers which represent their molecular weights, their masses, their densities, the temperature of the system, the pressure that it supports, *etc.*. Perhaps it is that in the body of the concrete system, taken under the experimental conditions that require certain values of the temperature, the pressure, a certain composition of the gaseous mixture, the water is indecomposable by all known means; I do not the less have the right, within the body of the schematic system, to decrease the numerical value of the mass attributed to the symbol H_2O and to increase in proportion the numerical values of the masses attributed to the symbols H and O; the operation has no physical meaning; but it does not contradict the abstract notions of the symbols H, O, H_2O, nor the definitions of the various quantities that characterise them; it constitutes a virtual modification.

CHAPTER III

EQUILIBRIUM AND MOTION

The notion of virtual modification was at the root of the Mechanics of Lagrange as it is at the base of the New Mechanics, but how more general it is in the latter than in the former! The only virtual alterations that were known to the Mechanics of Lagrange were the alterations of shape and position of the various parts of the system; many other alterations are considered by the New Mechanics.

An extension equal to this that has taken the notion of virtual modification affects the notion of real modification, or, as we shall say henceforth, of *motion*.

The only motion known to the Old Mechanics was the motion under which a body occupied different locations at different instants, *local motion*, to speak as did the peripatetic philosophers. The New Mechanics is not going to be restricted to the study of local motion; it will study also other kinds of motions whose variety will render to the idea of motion the huge extension that Aristotle recognised[1].

Without doubt, it will treat local motion, alterations in position and shape. But it will also treat alterations under which the various quantities of a body augment or diminish in intensity, under which a body is heated or cooled, magnetised or demagnetised. It will treat also those alterations of physical state by which a complete set of qualitative or quantitative properties is destroyed in order to give way for another set of quite different properties; such are the melting of ice, the vaporisation of water, the transformation of white phosphorus into red phosphorus. For it, these alterations will be motions; the Scholastic would have called them *motions of alteration*.

The examination of such motions will not fill yet all the domain that the New Mechanics pretends to submit to its laws; it

---

[1] See: Part One, Chapter I, *Peripatetic Mechanics*.

also comprises treating alterations where a set of substances disappears to let appear another set of substances, and such alterations the Peripatetics would have considered as *corruptions* and *generations*, and that today we would call *chemical reactions*. The New Mechanics is not content with being a *Physical Mechanics*, it is furthermore a *Chemical Mechanics*.

The extension taken by the idea of motion necessitates an equal extension of its opposite, the idea of *equilibrium*. A system in equilibrium will not only be a system which suffers no alteration of configuration or position; this will again be a system of which the various parts are neither heated nor cooled, upon which the electric and magnetic distributions remain unvaried, which suffers neither melting nor freezing, nor vaporisation, within the body of which no chemical reaction is produced. One will also speak not only of the equilibrium of configuration, but also of thermal, electric, chemical equilibria. Thus generalised, the notion of equilibrium will be the object of the New *Statics*.

From this Mechanics, which is the study of the equilibrium and motion extended to the broad meaning of Aristotle, we have defined the spirit and delimited the field of discussion; in gross characteristics we are going to describe its development.

CHAPTER IV

THE CONSERVATION OF ENERGY

   The New Mechanics is organised, not for the speculative and metaphysical contemplation of the essence of things, but for the practical necessity of acting upon the bodies of the external world and modifying them according to our needs. This character is first of all affirmed in the method it follows for posing its first principle, the *Principle of the Conservation of Energy*.
   At the heart of a material system, we can, by our efforts, produce a certain modification or assist this modification; we can displace a body, launch it with a certain velocity, deform it, break it, grind it; upon rubbing it we can heat it or electrify it. We can, against this, employ our efforts to put some obstacle in the way of the transformation that a system suffers; we can arrest a body in motion, slow it, stop it from being deformed. We then say that we have done a certain task, *accomplished a certain work*. The psychic and physiological intermediaries by which the efforts of our activity have produced a modification in the external world remain more or less hidden from our intelligence, but the effect that these efforts have produced is clearly perceived by our senses.
   Everyday experience teaches us that for our personal action we can substitute a body or an assemblage of bodies capable of producing or assisting the modification that we produce or that we assist, to hamper the modification that we hamper. Thus, through the course of the centuries, man has substituted for his action first of all the action of his peers, then that of animals, then that of inanimate machines more and more complex. Instead of grinding grain himself with a pestil and mortar, he had the mill turned by slaves, then by animals; finally, he used the wind or water mill. Instead of raising a load by the strength of his arms, he tied some cattle to a cord wound around a pulley block, then used a steam or hydraulic crane. Instead of throwing a projectile by hand, he used the tension in a string and then the explosion of gun powder.

CONSERVATION OF ENERGY

The first object of Mechanics is precisely this: to know what are the various bodies that can be substituted for our personal activity in order to promote or hinder a modification, what are the machines that can replace labourers for the execution of a certain task. The work that we would have had to carry out if we had acted ourselves upon the system that is transformed by the body or set of bodies which we have substituted for ourselves or our peers.

This notion of work accomplished by bodies external to a system whilst that system suffers a certain modification, let us transport it even to the case where the modification suffered by the system is of such a nature that our personal action can neither assist it nor hamper it—such as a chemical reaction. The work carried out by these external bodies is supposed to represent the work that an operator constructed in a different fashion from us would carry out and who would be capable of bringing to the system the help or hindrance that the external bodies would.

To summarise, when a material system is transformed in the presence of external bodies, we shall consider these external bodies as contributors to this transformation, either in stimulating it, or in helping it, or in impeding it; it is this contribution which we shall name *the work carried out, during a modification of a system, by the bodies external to the system*.

What is the nature of this contribution and how is it carried out? A difficult problem, whose clear solution seems to be quite beyond the bounds of human reasoning. But this problem of the *communication of substances* is the object of Metaphysics, not of Physics. Physics does not attempt to elucidate it; more modest, it endeavours solely to create a proper mathematical expression to serve as a symbol for this contribution, for this *work*; and this expression it seeks to construct with elements drawn from the effect that produces this action of the external bodies, for, if the nature of the action is obscure, the effect is clear and susceptible of observation.

To construct this mathematical symbol for the work that the bodies external to a system carry out in a modification of that system, one will not use arguments; what principles will serve them as the major ones? One will be guided by what one can, in the etymological sense of the word, call *inductions*; one will note the characters that it is most natural to attribute to this idea of work, and one will seek to express in the mathematical symbol analogous characters. Without following the details[1] of these inductions, let us note their essential steps.

One is led, first of all, to represent the work that the bodies external to a system carry out during a modification of this system by the increase suffered, under this modification, by a

---

[1] The reader will be able, if he so desires, to find this detail in our: Commentaire aux principes de la Thermodynamique, 1re partie, chapitre II, *Journal de Mathématiques pures et appliqués*, 4e série, vol. VIII, (1892).

certain quantity absolutely independent of the nature of the external bodies; this quantity is the *Total Energy* of the system.

The total energy depends upon two kinds of elements, both, it is well understood, proper to the system which is transformed and without any link with the external bodies which influence that transformation.

The elements that appear in the first place in the total energy are the numbers which measure, or reference the quantitative or qualitative properties of the system; the set of these numbers defines what is called the *state* of the system.

The elements which appear in the second place in the total energy are represented by the magnitude and the direction of the velocity that animates each point of the system, consequently its local motion; the former elements determine the *local motion* of the system at the instant considered.

A very simple assumption divides into two parts the work carried out in the course of a modification; one of these terms depends upon the alteration brought about in the state of the system; it depends neither upon the local motoin nor upon the alteratoin that this motion suffers; the other terms does depend upon the local motion and its alteration, but not upon the state of the system, nor upon its variations. From there the total energy is divided itself into two terms; the one depends only upon the state of the system and not upon its local motion; the other depends only upon the local motion and not upon the state. The first term would have to be called strictly the *energy of the state*; it is called the *internal energy* or the *potential energy*. The second term is called the *kinetic energy* or the *actual energy*.

In each of the chapters of Physics one determines by particular hypotheses the form that it is convenient to attribute to the internal energy of the systems studied in that chapter; the form of the kinetic energy, on the contrary, is susceptible of a general determination.

First of all it is easy to see what is the sum of the kinetic energies of each of the infinitesimal material elements into which the system can be assumed to be divided; and this remark especially simplifies its determination.

Let us take two different material elements and, starting from rest, let us project them with the same velocity; we carry out, in general, two different amounts of work; it is natural to think that the ratio of these two works is independent of the common velocity imparted to the two elements; this ratio, which depends solely upon the nature of these two elements, is called the ratio of the *masses* of the two elements. The mass of a material element is therefore proportional to the work that it is necessary to carry out to project it with a specified velocity.

The notion of mass being thus introduced in a very natural form, one sees that the kinetic energy of an element is the product of the mass of this element with a function of its velocity, this function being the same for all conceivable material elements. The determination of this function will be accomplished in render-

CONSERVATION OF ENERGY

ing explicit the expression for the kinetic energy.

One could attempt — this would be the most obvious hypothesis — to take this function as simply proportion to the velocity; this attempt would lead to constructing a Mechanics which, in its essential characteristics, would reproduce Cartesian Dynamics. The memorable failure of this Dynamics gives us notice not to set upon this path. It is then natural to take again Leibniz's idea and to regard the unknown function as being proportional to the square of the velocity. The kinetic energy, completely specified, becomes identical with that which the Old Mechanics named the *live force*.

Let us now take an isolated system; there exists no body that is external to the system; therefore, under every modification of this system the work accomplished by the external bodies is nil; in other words, *under every modification of an isolated system, the total energy of this system keeps an invariant value*. We are here in possession of the first principle of the New Mechanics, the *Principle of the Conservation of Energy*.

The enunciation of this principle offends those who wish to see in the axioms of Mechanics generalised experimental laws, who wish to attribute a physical meaning to the first postulates; for, in Nature, there exists no isolated system. For us such an objection contains nothing embarrassing; we know that the principles of Theoretical Physics are simply rules by which we impose a specified form upon the mathematical scheme we wish to construct; it is not necessary for these postulates to have a physical meaning; only do their later consequences have to accord with facts. Now, as long as we try no comparison with the external world, as long as we keep within the domain of the abstract mathematical scheme, we imagine perfectly that a system contains all bodies studied, that there exist none outside it, that it is isolated in pure space.

CHAPTER V

WORK AND QUANTITY OF HEAT

Always, in order that our mathematical scheme does not remain sterile, is it necessary for us to extend it and in no way limit it to the consideration of a system isolated in space.

If we take, on the one hand, the material system whose modifications we wish to study, and, on the other hand, all the bodies whose presence does not seem indifferent to these modifications, we can treat the set of these two mutually independent systems as comprising a single system isolated in space; to this isolated system we can apply the principle of the conservation of energy.

The live force of this complex system is the sum of the live forces of the two independent systems that compose it; but the internal energy of the complex system is not equal to the sum of the internal energies that the two independent systems would have if each of them were isolated in space; it is equal to this sum augmented by a term which can be called the *mutual energy of the two systems*[1].

The existence of this mutual energy signifies that the properties of each of these two systems, taken in the presence of the other, are not the same as if this system existed alone in the same state; that the presence of each of them is not indifferent to the other.

This mutual energy depends upon the state and upon the position of the first system, that is to say upon independent variables $\alpha, \beta, \ldots$ which determine this state and this position; it also depends upon the state and the position of the second system, that is to say upon independent variables $\alpha', \beta', \ldots$ which determine this state and this position.

Let us imagine that a *virtual* modification now affects the set of our two systems, imposing upon the variables $\alpha, \beta, \ldots$ some in-

---

[1] *Translator's Note:* Nowadays one thinks of *interaction energy*.

finitesimal variations $\delta\alpha, \delta\beta, \ldots$ and upon the variables $\alpha', \beta', \ldots$ some infinitesimal variations $\delta\alpha', \delta\beta', \ldots$. The mutual energy of the two systems suffers a diminution which is found to be the sum of two terms; the first of these terms is of the form $A\delta\alpha + B\delta\beta + \cdots$; the second is of the form $A'\delta\alpha' + B'\delta\beta' + \cdots$; the quantities $A, B, \ldots, A', B', \ldots$ depend upon the state of our two systems and their mutual position.

The sum $A\delta\alpha + B\delta\beta + \cdots$ is what we call the *virtual work of the actions exerted by the external bodies* upon the system studied; similarly, the sum $A'\delta\alpha' + B'\delta\beta' + \cdots$ is the virtual work of the actions that the system studied exerts upon the external bodies.

Let us dwell for a moment upon these notions which, in the development of the New Mechanics, will play a capital rôle.

The systems that the Old Mechanics treated were defined entirely by their form and their position; the variables which determine the state of the similar systems are exclusively geometric. Let us apply the preceding considerations to A in the similar systems: in the sense of Lagrange's Mechanics, the sum $A\delta\alpha + B\delta\beta + \cdots$ will become the virtual work of all the *external forces* applied to the system; A will be the generalised external force which corresponds to the independent variable α; if α represents a length, A will be a force, in the elementary sense of the word; if α represents an angle, A will be the moment of a force.

The properties of the systems that we study now are no longer entirely reducible to the configuration and to the position of their various parts; amongst the variables which define the state of these systems, there are some that represent neither lengths, nor angles, nor surfaces, nor volumes, nor anything geometrical, but physical quantities, temperatures, electric charges, intensities of magnetisation. If α represents such a variable, A will no longer be a generalised force, in the sense of Lagrange's Mechanics; it will be a quantity of quite another nature, having only this character common with the generalised force; its product with the infinitesimal variation of the variable A represents a work. If, for example, α is a magnetic moment, A will be the component in the direction of this moment, of the external magnetic field. Of such a quantity we shall say that it represents the *external influence* relative to the physical variable α, and we shall reunite the generalised forces and the influences under the common name of *actions*.

Here, as in Lagrange's Mechanics, the actions that some specified external bodies exert upon a system also specified are not entirely defined quantities; they change if one changes the group of variables that serves to represent the state of the system; alone does the work that they perform under a specified virtual modification keep an invariant value.

The virtual modification, which has already furnished us with the definition of the external actions exercised upon a medium, is going to furnish us with another essential notion, that of the *quantity of heat* that the system releases in such a modification.

We shall reach this new notion by applying the principle of the conservation of energy to our virtual modification.

What will have to be understood by that?

The statement of the principle of the conservation of energy brings into play the increase in the kinetic energy of the system; this increase has a meaning only for a real modification; a virtual modification does not bring it about in time; it does not communicate to the kinetic energy of the system any change; it is enough to say that, in its primitive form, the principle of the conservation of energy does not apply to virtual modifications. We are free, it is true, to impose upon it a generalisation that renders it applicable to these modifications and this freedom is only limited by a single condition: The new principle will have to reproduce the first one when for a virtual modification one substitutes a real modification.

Now, we know that under a real modification the work of the inertial forces is equal to the diminution of the kinetic energy. If, therefore, we take a proposition which has the characteristic of real modifications alone because it contains the words *decrease in kinetic energy*; if for those words we substitute these: *work by inertial forces*, we obtain a new statement applicable to virtual modifications, and which contains, as a particular case, the primitive statement.

It is by this process that we shall understand the principle of the conservation of energy for virtual modifications and which we shall express by the following proposition: *In every real or virtual modification of an isolated system, the work performed by the inertial forces is equal to the increase in the internal energy*.

Let us take our isolated system formed by the union of two independent systems and, for a virtual modification imposed upon this system, let us calculate the sum of the work of the inertial forces and the decrease of the internal energy, a sum whose value has to be zero. This sum will contain six terms, of which the first three are:

(1) The work of the inertial forces applied to the first system;
(2) The work of the actions exerted by the second system upon the first system;
(3) Finally, the decrease in the internal energy of this first system.

The last three terms, analogous to the three first ones, are deduced from them by inverting the rôle of the first system and the second.

The sum of these six terms is zero; but in general this is not so for either the sum of the first three nor the sum of the last three.

The sum of the first three is, by definition, the *quantity of heat* that the first system releases in the course of the modification under consideration; the sum of the three last terms is the quantity of heat released under the same modification by the second system; these two quantities are equal and of opposite sign.

By this definition, when a system suffers an arbitrary virtual modification, *the virtual work of the external actions augmented by the virtual work of the inertial forces gives a sum equal to the increase of the internal energy of the system augmented by the quantity of heat that this system releases.* If it is a question of a real modification, this proposition is transformed into the following: *The increase in the total energy of a system is equal to the excess of the work performed by the external actions over the quantity of heat released by the system.*

This proposition is the exact statement of the law of the equivalence between heat and work[1]. This law appears here for us as a corollary of the principle of the conservation of energy coupled with the definition of work and quantity of heat.

This completely algebraic definition of the quantity of heat might perhaps horrify some people; they would be astonished to see the words *quantity of heat* used to denote a sum of terms in whose construction the notions of hot and cold are completely foreign. Their astonishment would have some reason to it, for the expression *quantity of heat*, imposed by usage, is a very badly chosen term, and one very capable of engendering deplorable confusion; the history of Physics makes that quite clear.

But if the preceding definition springs before our eyes before our eyes the absence of every logical link between the notion of *quantity of heat*, as understood by the physicist, and the notion which comes to us through our perceptions and that is expressed in vulgar language by the employment of the word heat, it is not this definition that has broken this link; it was broken at the origin of experimental physics on the day that the Academicians of Florence proved that in *heating* ice it melted *without heating it*. From this day dates the distinction between *temperature*, the translation into physical language of the empirical notions of hot and cold, and *quantity of heat*; the calorimetric investigations of Black, of Crawford, of Lavoisier and of Laplace, the concept of latent, have only served, daily, to make this separation more deeply entrenched.

It is therefore correct that the definition of quantity of heat borrows nothing from perceptions of hot and cold; but it would be inadmissible that the quantity so defined should remain without a relation with that which physicists measure by means of the calorimeter. This relation, happily, is established without difficulty[2]; the principles that we have just expounded prove

---

[1] See: Part One, Chapter X: *The Mechanical Theory of Heat.*
[2] One will find the establishment of this relation, as well as the mathematical development of the present Chapter, in our: Commentaire aux Principes de la Thermodynamique, 1re partie, chapitre III, *Journal de Mathematiques pures et appliqués*, 4e série, vol. III, (1892).

that the calorimeter measures effectively what we have named
quantity of heat; the definition of this quantity therefore satisfies the rule posited by Rankine; it has as an almost immediate
corollary a process proper to measuring the quantity defined.

The two notions of work and quantity of heat are continually
at play in the New Mechanics whose development we are sketching;
one can therefore very correctly name this Mechanics *Thermodynamics*; with Rankine one can also give it the name of *Energetics*,
for the notion of energy is the source from which it flows in
its entirety; between the supporters of the first name and the
followers of the second we shall not attempt to make a settlement:
"Simus faciles in verbis", as Gauss said.

CHAPTER VI

THE REVERSIBLE MODIFICATION

Up to now we have treated properties of systems studied without having to distinguish between them; all have played the same rôle; the letters $\alpha, \beta, \ldots$, which denoted the variable quantities by which these properties are represented, could as well represent lengths, or angles as temperatures or intensities of magnetisations.

It is a number, a symbol of a physical quantity, the *temperature*, which henceforward is going to play a separate and quite exceptional rôle; this rôle is going to be attributed to it by the principle that Sadi Carnot discovered, that Clausius modified and perfected, and which is one of the foundations of the New Mechanics.

The statement of this principle will use this strange expression: A *reversible modification*; this expression denotes one of the most delicate notions of Thermodynamics; before all else it is necessary, therefore, to analyse this notion.

The transformations which are actually produced in the real world are never reversible.

Here is a gas contained within the body of a pump that a piston seals; this piston is loaded by weights. If the load is large enough the piston is going to advance and the gas will be condensed; the external actions, represented here by the weights that load the piston, will perform a positive work; a certain quantity of heat will be released. If, in the contrary case, the weight that loads the piston is too small, the piston is going to rise; the gas will expand; the work of the external forces will be negative; the system will absorb heat. It can be arranged so that one obtains the first group of phenomena or that one obtains the second group. But to seek to place upon this piston a weight such that without changing this weight the piston may just as well de depressed as raised; that the gas may be either condensed or expanded; that it may at will have a release or absorption of heat, this is evidently a chimerical task. A given system, placed in conditions also given, is necessarily transformed in a determined sense; it is not transformed indifferently in one sense or in the opposite sense; taken literally, the words *reversible modification* are a non-sense.

These words, however, are capable of taking a precise meaning, but by a detour that we shall have to follow.

Upon loading with a suitable weight the piston which compresses a gas, we can cause the piston to advance; with a load a little less, it could still be advanced; in order that the piston begin to be advanced it suffices that the load exceed by just a little that weight which would hold the gas exactly in equilibrium; similarly, in order that the piston rise, it suffices that it carry a charge just a little less than that which would hold the gas in equilibrium. We can therefore take two charges which will differ from each other as little as one may wish and choose them in such a way that the one will oblige the piston to advance and that the other will let it raise; between these two charges there is that which assures us of the piston's immobility.

A given system, encompassed by circumstances also given, suffers a modification whose sense is always perfectly specified; but one can choose the external conditions in such a way that an infinitesimal variation of these conditions will suffice to reverse the sense of the alteration of state that thay determine; it is necessary, for that, that the bodies by which the system is surrounded differ infinitesimally from those which would maintain it in equilibrium.

What is it then, to be definitive, for a *reversible modification* to be suffered by a system? It is a purely ideal, purely virtual modification, a *continuous sequence of equilibrium states* in each of which the system is successively conceived by the reasoning of physicists; and *this sequence of equilibrium states is the common boundary of the series of real modifications, of which the one proceeds in a certain sense and the other in the opposite sense*.

The abstract systems to which Physics has recourse to represent the world of inert matter are not all, and it is necessarily so, susceptible to reversible modifications.

A metal wire, tautened by weights, is in equilibrium; now let us increase the tensioning weight; the wire is stretched with a certain speed and arrives at a new state of equilibrium; a new enloadening produces a new stretching; and so on. On a table let us mark the sequence of loads used, and with each the lengths that the wire took when in equilibrium under each of these loads.

Let us recommence this sequence of experiments starting from the same initial state, but by using, in each operation, an enloadening less than that in the previous case. We shall obtain a new table where there will appear more numerous equilibrium states than in the first, and closer together.

Let us take yet a third and a fourth series, with enloadenings successively smaller and smaller; the tables obtained will tend towards a certain limiting table; the latter, if it were possible to form it, would present a sequence of loads increasing continuously and, accompanying it, a sequence of lengths increasing continuously also; each of the lengths would be that of the wire when it holds in equilibrium the load placed opposite it. We would thus have obtained a continuous sequence of equilibrium states, and this continuous sequence, *run through in the sense in which the lengths increase*, would be the limit form of a series of experiments in the course of which the wire does lengthen in reality.

Now let us take the wire in the last of the equilibrium states to which the previous trials have led, and, gradually unloading, let it shorten until it takes again its initial length. The more the successive decreases of the load are small, the more slowly will be the contraction of the wire. It will then be possible for us, by operation as for the lengthenings, to form a continuous sequence of equilibrium states of the wire, and this sequence, *run through in the sense in which the lengths decrease*, will represent the limit form of a series of real contractions.

Let us compare the two sequences of equilibrium states thus constituted; they are not identical to each other; to one load there corresponds, in the second sequence, a length of the wire greater than in the first, and this is expressed by saying that the stretching of the wire has produced in the wire a permanent lengthening; our two series of real modifications, in opposite senses, the stretchings on the one hand and on the other hand the contractions, do not admit a common frontier; a wire susceptible of permanent lengthening cannot be subjected to a reversible modification.

The Mechanics which we are going to develop will make a continual use of the notion of reversible modification; *it will exclusively treat systems for which every continuous sequence of equilibrium states is a reversible modification*; by doing so it will cease to be an entirely general Mechanics in order no longer to be the study of a category, doubtless very extended, but nevertheless particular, of material systems; outside of the domain that it pretends to submit to its laws, it will leave plenty of bodies, namely those which can suffer permanent modifications; if, later, a Mechanics can be constructed which embraces in its theorems the equilibrium and the motion of similar bodies[1], it will be by a generalisation of the restricted Mechanics which is now going to occupy us, by a contribution of hypotheses and principles foreign to those which we are going to announce.

In the restricted domain within which we are going to billet ourselves, what will be the use of this purely ideal and fictitious notion that the words reversible modification express? What exactly does this phrase mean: Such a proposition is true only for a reversible alteration? The meaning of this phrase is this: Properly speaking, the proposition to which it refers

---

[1] A similar Mechanics will be studied in Chapter XIV.

is never true; there exists no real modification to which one can rigorously apply it; but the error is committed by applying this proposition to an alteration able to be more or less large; it is just as small as, to reverse the sense of this alteration of state, would be necessary to impose a smaller perturbation upon the external conditions which surround the system subjected to the alteration; the proposition in question is just as little removed from the truth as the actions to which the system is submitted are, at each instant, closer to those which maintain it in equilibrium.

The Carnot-Clausius Principle is only true for reversible modifications; the consequences we shall deduce from this principle, the properties that it will lead us to discover in a system, will never be rigorously exact as long as the system is in the process of transformation; but the more the causes which determine this transformation may tend to disappear, the more these consequences will be close to those that experiment reveals; for systems in equilibrium the propositions will apply fully. *The notion of reversible modification is able to serve for founding a Statics.*

CHAPTER VII

CARNOT'S PRINCIPLE

AND ABSOLUTE TEMPERATURE

If the Principle of the Conservation of Energy can be prescribed by the instinct which compels us to act upon the external world and to modify it according to our needs, *a fortiori* is this similarly so for Sadi Carnot's Principle. It is an historic fact that this Principle was suggested to its author by the contemplation of heat engines and by the ambition of giving a completely general theory. It is, in particular, this contemplation which led Sadi Carnot to imagine the sequence of operations that today one calls the *Carnot cycle*.

A system describes a *cycle* when it suffers a sequence of operations which lead it back to its initial state; if all these operations are reversible, the cycle itself is reversible. In the course of a cycle the system can now release, now absorb heat. Let us assume that these exchanges of heat between the system and the external bodies takes place only under two circumstances: firstly: when all the bodies which comprise the system are taken to a certain temperature ϑ, secondly, when all these bodies are taken to a certain temperature ϑ' greater than ϑ. The cycle will be a Carnot cycle described between the two temperature limits ϑ and ϑ'.

Of the hypothesis formulated by Sadi Carnot we shall say nothing here; they do not agree with the Principle of the Conservation of Energy; also Clausius, then Kelvin, modified them and formulated two postulates that are universally accepted.

The *Clausius Postulate* can be enunciated in the following way: *In order that a system describing a reversible Carnot cycle absorb heat whilst it is carried to the lower of two temperature limits, it is necessary that the external actions to which it is subject carry out a positive total work during the course of the cycle.*

The *Kelvin Postulate* has a similar form; here it is: *If the external actions which act upon a system carry out a negative total work during the course of a reversible Carnot cycle, the system necessarily releases heat during its traverse from the lower temperature limit.*

From these two postulates is deduced[1] a set of consequences which forms *Carnot's Theorem*.

When a system describes a reversible Carnot cycle, the quantity of heat that it releases whilst its temperature reaches one of the two limits is of sign opposite to the quantity of heat that it releases whilst it is carried to the other temperature limit; if, in the first case, it releases heat, it releases it in the second, and conversely. The absolute values of the quantities of heat brought into play have a certain ratio between them; the value of this ratio depends neither upon the nature of the bodies which describe the Carnot cycle, nor upon the particular form of the modifications which compose this cycle, nor even upon the thermometric scale upon which are read the two temperatures ϑ, ϑ'; it depends exclusively upon the two *intensities of heat* to which, by the choice of an appropriate thermometer, the two numbers ϑ, ϑ' have been made to correspond; if one changes this thermometer, the numerical values of the temperatures which correspond to the same intensities of heat will be changed, but the value of the ratio considered will remain invariable.

In other words, to each intensity of heat one can bring into correspondence a number; this number is always positive; it is the larger as the quality of heat to which it corresponds is the more intense, a character which permits this number to be taken as the temperature, to regard the sequence of numbers so defined as a thermometric scale; the correspondence between each of these numbers and the intensity of heat that it can serve to reference is in no way related to the choice of a particular thermometer, so that the temperature thus defined merits the name of *absolute temperature*; the employment of this terminology allows one to formulate the proposition announced previously in the form here: *When a system describes a reversible Carnot cycle, the absolute values of the quantities of heat that it releases or absorbs whilst it attains one or the other of the limiting intensities of heat are in the same proportion to each other as are the absolute temperatures which correspond to the intensities of heat.*

One last proposition accomplishes the sharpest statement of this so essential a notion of absolute temperature. It requires the consideration of those fluids that physicists call *perfect gases*, and which they define by two characteristics: a compressibility which, at a constant temperature, obeys Mariotte's law[2], an internal energy which remains constant as long as the gas also remains cold. Carnot's Theorem, in fact, implies this consequence:

---

[1] This deduction is expounded in the majority of recent Treatises on Physics; we think to have given it a completely rigorous form in our: *Traité élémentaire de Mécanique chimique fondée sur la Thermodynamique, Tome I*, (Paris), (1897), Livre I, chapitre III, p. 56.

[2] *Editor's Note:* Also known as the Townley-Boyle-Mariotte Law.

One can take for absolute temperature the Centigrade temperature read on a perfect gas thermometer augmented by the inverse of the expansion coefficient of this gas.

This proposition completes the definition of absolute temperature in conformity with the law proposed by Rankine: it traces for us, in fact, a sketch of a process which will permit the measurement of absolute temperatures. There does not exist in Nature a perfect gase which we might introduce into a reservoir to turn it into a thermometer; the perfect gas is a concept constructed entirely from pieces of our reasoning; it has no more concrete a reality than the perfectly undeformable solid treated by Elementary Mechanics. But if there does not exist, in Nature, a rigid solid, there exist bodies that are deformed very little when the temperature and external actions do not exceed certain limits; to these bodies the propositions of Elementary Mechanics apply approximately, and on this side of the limits which we have must mentioned. Similarly, concrete reality presents us with no perfect gas; but certain real gases, provided that they are not too compressed, nor too cooled, may be approximately represented by this scheme, a simple grouping of mathematical elements, which the words *perfect gas* denotes. With this latter gas, one will be able to construct thermometers which give the absolute temperature. The determination of the absolute temperatures will only be possible if the experimental conditions remain bound within certain limits; between these limits it will only be approached; these characters are common to all the measuring processes used in Physics.

CHAPTER VIII

INTERNAL POTENTIAL

AND GENERAL STATICS

We have seen that the study of reversible modifications can serve to establish the propositions of Statics; in fact, from Carnot's Theorem one can draw the general properties of systems in equilibrium.

In this study there is great interest in making use, for representing the properties of the system, of certain special variables that one calls *normal variables*. The temperature always appears in the number of normal variables, but it appears there in a particular rôle; these variables, in fact, are chosen so that under a virtual modification where the temperature alone changes, whilst each of the other variables keeps its value, the various parts of the system remain still and the external actions do no work.

One will wonder, no doubt, if the properties of any system can be represented by such variables; assuredly no; a fluid would be expanded under a raising of temperature but whose compressibility would be zero at constant temperature could not be defined by normal variables; but, practically speaking, all the systems that physicists are led to conceive of for representing the properties of bodies can be referred to normal variables.

The use of normal variables gives to the propositions of Thermodynamics their simplest form; henceforward we shall adopt this usage.

The Principle of the Equivalence of Heat and Work and Carnot's Principle then lead to capital consequences which we are going to inspect on review.

To each state of the system that is studied, these two principles attach a certain quantity, specified when one knows the absolute temperature of the system and the other normal variables which determine its properties. The consideration of the quantity dominates the whole of Thermodynamics. F. Massieu, who first brought it to the attention of physicists, called it the *Charac-*

teristic Function of the system; for Gibbs and Maxwell it was the *Available Energy* (*l'Energie utilisable*), for Helmholtz the *Free Energy*; we have given it the name of *Internal Thermodynamical Potential*. The multiplicity of these titles has a reason, for each of them reflects one of the aspects under which one can consider this quantity; they will all find their justification in the developments which are going to follow.

From the expression of this quantity one draws, without difficulty, the necessary and sufficient conditions for the system to be maintained in equilibrium by some external bodies kept at the same temperature as itself.

In order to obtain these conditions one imposes upon the system a virtual modification *that does not change its temperature*; to this modification there corresponds a certain virtual work of the external actions and a certain increase of the internal potential; one expresses the mutual equality of this virtual work and this increase.

The fundamental principle of the New Statics is therefore presented exactly in the form that Lagrange[1] gave to the principle of the Old Statics; the quantity whose existence has been revealed to us by the axioms of Thermodynamics plays in the former the rôle that the potential of the internal forces played in the latter; from this there comes the name of Internal Thermodynamical Potential which we have attributed to this quantity.

The profound analogy between the fundamental principles implies intimate comparisons between the two sciences that flow from them; we are also going to rediscover, in the domain of the General Statics, all the very fertile ideas that Lagrange created whilst expounding a more restricted Statics.

The formation of the equations of equilibrium readhes its highest degree of simplicity in the case where the state of the system studied is entirely defined by the absolute temperature and by the normal variables which are mutually independent; in this case these conditions are stated in the following manner: Each of the external actions $A, B, \ldots, L$ which correspond respectively with the normal variables $\alpha, \beta, \ldots, \lambda$, other than the absolute temperature, is equal to the derivative of the internal potential $\mathcal{F}$ with respect to the corresponding variable. For this statement one can substitute the following equations:

$$A = \frac{\partial \mathcal{F}}{\partial \alpha}, \quad B = \frac{\partial \mathcal{F}}{\partial \beta}, \quad \ldots, \quad L = \frac{\partial \mathcal{F}}{\partial \lambda}. \qquad (1)$$

The number of these equations—and this remark will be of some consequence—is equal to the number of normal variables that it is necessary to adjoin with the temperature in order to determine the state of the system entirely; they fix the value of each of these variables and, consequently, the equilibrium state of the

---

[1] See: Part One, Chapter VI: *The Principle of Virtual Velocities and Lagrange's Statics*.

system when the temperature and the external actions are given.

It may occur that the variable quantities by which one represents the properties of the system are not mutually independent, that they are rendered interdependent by certain constraint conditions; then shall we recover, in constructing our arguments upon those of Lagrange, the constraint forces, however, the generalised constraint forces, as have the forces themselves, have become *constraint actions*.

The spirit of the methods of Lagrange's Statics has therefore passed in its entirety into the general Statics, the conception of which will forever redound to the glory of J. Willard Gibbs[1]; but in passing from the one to the other they have evolved; the germs sown by the author of *Analytical Mechanics* owe their ample and full development to the physicist who treated the *Equilibrium of Heterogeneous Substances*.

Let us cast our eyes around the science that has issued from this development.

In every field it exceeded the limits of the Old Statics.

From the study of compressible fluids this science is found to be reduced to confessing its own insufficiency. Amongst the equilibrium conditions of these fluids its brings into play a relation between the density, the temperature and the pressure; this relation could not be drawn from its own principles; it was introduced straightway as a postulate suggested by experiment. In defining compressible fluid as a medium whose every element is in a known state when the density and temperature are known, the New Statics can form the expression for the internal potential of such a fluid and discuss the equilibrium conditions[2]. These conditions are much more general than the Hydrostatics of Clairaut, Euler and Lagrange ever assumed; in particular, the existence of a relation between the pressure, density and the temperature could only be given as a completely general rule; it is proper to the bodies that form a particularly simple category amongst all possible fluids; happily, this category embraces the largest number of cases met in practice.

When several fluids interpenetrate and mix together, the whole remaining subject to external forces, to the weight, for example, they are distributed according to laws that escaped the efforts of Lagrange's methods; in his immortal writing *On the Equilibrium of Heterogeneous Substances*, J. Willard Gibbs drew these laws from the principles of the New Statics; thus he was able to give, for the effects of osmosis, a theory whose principal propositions are in current usage today.

The Statics founded upon Thermodynamics can, as Kelvin has shown, render analogous services to the theory of elastic equi-

---

[1] J. Willard Gibbs: On the Equilibrium of heterogeneous Substances, *Transactions of the Academy of Connecticut*, vol. III, (1875-1878).

[2] P. Duhem: Le Potentiel thermodynamique et la Pression hydrostatique, *Annales de l'École Normale Supérieure, 3e série*, vol. X, (1893), p. 183.

librium; but the fecundity of it is manifested still better in the study of purely qualitative properties such as electricity and magnetism.

To draw from Rational Mechanics the laws of electrical and magnetic equilibrium, Poisson was obliged to regard electricity and magnetism as fluids and to make various hypotheses about the properties of these fluids. The collapse of the doctrine of caloric implied the discrediting of electric and magnetic fluids. One then asked for special postulates, the ones suggested by experiment, the others conceived of *a priori*, the laws which governed a permanent distribution of electricity or magnetism. This method had permitted the reduction to mathematical analysis of a great number of problems belonging to this branch of Statics; but it established no logical link between the hypotheses upon which the various solutions rested.

Some of the hypotheses formerly admitted suffice to form the internal potential of a system in which there appear electrified bodies, polarised dielectrics and magnets; once this potential is known, the theory of electrical and magnetic equilibrium unfolds in its totality by regular calculations where indeterminacy no longer has a place; electrification of conductors, homogeneous or heterogeneous, whose temperature is uniform, and thermo-electric chains; magnetisation of isotropic or anisotropic bodies; polarisation of amorphous dielectrics, of holomorphic or hemimorphic crystals, all these calculations depend upon equations which a unique process supplies[1], traced over the method used in Statics by Lagrange.

It is first necessary, in order to put into equations a new problem of electrical or magnetic Statics, to have recourse to new postulates; the excessive freedom left to the physicist in the choice of these new hypotheses would only generate error and confusion when it came to a question of treating a new and complicated question; thus it was that the theory of the deformations that affect a solid or elastic fluid when that body carries a dielectric or magnetic polarisation received from Maxwell, from Helmholtz, from Korteweg, and from Kirchhoff, an unacceptable form; the processes of the New Statics have been able to unravel the complications of this problem[2].

The services rendered in the domain of electricity and magnetism would not perhaps have sufficed for assuring the triumph of the New Statics; in plenty of cases, in fact, the results to

---

[1] P. Duhem: *Théorie nouvelle de l'aimantation par influence, fondée sur la Thermodynamique*, (Paris), (1888); *Leçons sur l'Électricité et le Magnétisme*, vol.s I, II, (Paris), (1891-1892).

[2] P. Duhem: *Leçons sur l'Électricité et le Magnétisme*, vol. II, (Paris), (1892), livre XII; Liénard: Pressions a l'intérieur des aimants et des diélectriques, *La Lumière électrique*, vol. LII, (1894), pp. 7,67; P. Duhem: Sur les pressions dans les milieux diélectriques ou magnétiques, *American Journal of Mathematics*, vol. XVII, (1895), p. 117.

which it led were already known; without doubt, these results had
not been deduced from general principles, but from hypotheses
special to each problem; without doubt, in some circumstances
they offered obscurities and contradictions that the thermodynamical method had made to disappear; yet, the conquests of this
method did not have the striking and convincing character of invention.

Fortunately, since its beginnings, Thermodynamical Statics led
J.W. Gibbs to the discovery of new laws, the importance of which
is reaffirmed more and more clearly day after day. It was in
studying the alterations of physical state or chemical constitution that the illustrious American created these laws. No domain
had been more closed to the Old Mechanics, none was more foreign
to the theory of local motions, than the domains of *generation*
and *corruption*, as the Peripatetics had called them, than *Chemical Mechanics*, in modern language. The kinetic hypothesis, as
the molecular attraction hypothesis, had in vain attempted to or-
ganise Chemical Mechanics. With the first blow, the Static founded upon Thermodynamics gave its measure by imposing upon it rules
that were as simple as they were profound; all these rules are
dominated by the *Law of Phases*.

In a flask in which a reaction occurs or a chemical equilibrium is established, the observer sees, isolated from each other
various substances, each of which has at every point the same
nature and the same properties; these substances are the *phases*
into which the chemical system is divided; Icelandic spar, lime,
carbonic gas are the three phases of a system where calcium carbonate is dissociated into lime and carbon anhydride. The number
of phases into which a chemical system is divided is one of two
numbers that characterise this system; the other is the number of
independent components which constitute it, that is to say, the
number of bodies whose mass is left arbitrary by the chemical
formulae of the substances entering into reaction. It suffices
to know these two numbers in order to be able to indicate the
general form in which the system's law of equilibrium is cast.

In the work of Gibbs[1] this law of phases for a long time contained an unrecognised theorem of Algebra; van der Waals exhumed
it from its cemetry of equations and pointed it out to the experimentalists; Bakhuis Roozeboom, van't Hoff, and their numerous
disciples made use of it for discussing chemical reactions so
complicated that they were inextricable without this help. Thanks
to the activity of these chemists, the scope of this new law can
no longer be contested; it has been said, not without reason,
that it would exercise upon the Chemistry of the twentieth century an influence comparable to that which Lavoisier's law exercised upon the Chemistry of the nineteenth century. From then

---

[1] J.W. Gibbs: On the Equilibrium of heterogeneous Substances,
Transactions of the Academy of Connecticut, vol. III, (1876), p.
152; Translated by H. Le Chatelier, under the title: *Équilibre
des systèmes chimiques*, p. 63.

on, the Law of Phases has profoundly transformed the theory of isomorphism; it has unravelled the chaos that, up to now, formed the study of mixtures; it has overthrown the ideas that chemists adopted touching upon the tokens by which one recognises a specified compound[1]. Drawn from extremely simple and general hypotheses, the Law of Phases extends to the whole of Chemical Mechanics; but it does not reach to details of phenomena; the information that it gives is qualitative rather than quantitative. Upon particularising the hypotheses which determine the internal potential, one will obtain some consequences which will delve more intimately into the analysis of phenomena. It was by thus attributing the properties of perfect gases to all the bodies that enter into reaction or only with some of them, that Horstmann[2] and Gibbs[3] were able to obtain formulae which agree numerically with the results of investigations into dissociation.

Thermodynamics therefore extends to the most varied domains: static electricity, magnetostatics, chemical statics, the methods created by Lagrange to treat Statics purely mechanically; but this extension itself, however prodigiously ample it may be, does not exhaust the fertility of the new discipline; to the determination of the equilibrium conditions of various systems there now comes to be added a chapter of which the Old Mechanics could not even conceive the possibility: the determination of calorific properties.

The development of the whole of this new chapter rests essentially upon the following analytical fact: When the internal potential of a system is known, a regular and, in fact, very simple calculation yields the internal energy.

Now let us consider a virtual modification arising from a state of equilibrium; under this modification the inertial forces are all zero; the quantity of heat released is the excess of the virtual work of the external actions over the increase of the internal energy; but the external actions which the system in equilibrium suffers, as well as the internal energy, are known by the internal potential; hence, similarly so, is the quantity of heat released; from the expression for the internal potential a regular calculation will draw the *calorific coefficients of the system* in equilibrium.

Thus, the investigation of the equilibrium conditions of a system will always and immediately be able to be completed by the investigation of the calorific properties of this system; the second investigation will be the natural sequel of the first. For example, Statics apprises us that one maintains a liquid

---

[1] The reader will find an exposition of the set of chemical investigations to which the Phase Rule has given rise in our book: *Thermodynamique et Chimie, leçons élémentaires à l'usage des chimistes*, (Paris), (1902).

[2] Horstmann: Theorie der Dissociation, *Annalen der Chemie und Pharmacie*, vol. CLXX, (1873), p. 192.

[3] J.W. Gibbs: *loc. cit.*.

covered by its vapour in equilibrium by applying to the two fluids a pressure that depends upon the temperature alone; at once this information is completed by the expression for the latent heat of vaporisation and specific heats of the two saturated fluids; the equilibrium law that gives the Law of Phases is immediately accompanied by the Clapeyron and Clausius formulae.

It would take too long to enumerate here all the works that rely on this order of investigations; we shall indicate only one of them. The calorific laws of electrolytic phenomena for a long time have been a bone of contention; a very simple formula given by Helmholtz, by Joule, and by Kelvin, agrees in no way with the experimental determinations of P.-A. Favre, of Raoult, F. Braun; the new methods allowed Gibbs and Helmholtz to resolve this difficulty and to establish some formulae that experiment verifies minutely.

An infinitesimal reversible modification is nothing other than a virtual modification arising from an equilibrium state; the quantity of heat released in such a modification is therefore determined by starting from the internal potential. Let us divide this heat by the absolute temperature in order to obtain, for our reversible modification, what Clausius named the *transformation value*[1]; this transformation value is found to be the decrease that a certain quantity, the *Entropy*, suffers in the modification considered, and is completely fixed when one is given the state of the system.

If one causes the system to run through a reversible cycle, the Entropy retakes its initial value at the end of the cycle; zero represents, therefore, the sum of quotients that one obtains when one divides each of the infinitesimal quantities of heat released in the course of a reversible cycle by the absolute temperature of the system during this release.

Discovered by Clausius, these propositions went before the constitution of the New Statics; they stimulated its creation and presided over its birth; besides the internal energy, they introduced another function of the state of the system, the Entropy; today, these two fundamental functions yield place to the internal potential, from which they are derived by a regular calculation.

Hence, when one knows the internal potential of a system, one knows from it the equilibrium conditions, the internal energy, the entropy, the calorific coefficients; in a word, the static study of the system is accomplished; the characteristics of the system in equilibrium are nicely and completely engraved upon it. It is this that F. Massieu had first seen, and it is why he gave the name *characteristic function* to the quantity which, later, we called internal potential.

---

[1] See: Part One, Chapter X: *The Mechanical Theory of Heat*.

CHAPTER IX

THE PRINCIPLE OF GENERAL DYNAMICS

The study of a material system placed under conditions where it no longer suffers any modification, a system *in equilibrium*, is complete; it is now necessary for us to approach the study of a system whose state changes from one instant to the other, of a system *in motion*, this latter word being taken in the larger sense that Peripatetic Physics attributes to it.

To pass from the laws of equilibrium to the laws of motion, the process that first offers itself to the physicist consists in extending to the General Mechanics the classical d'Alembert's Principle[1].

By virtue of this principle, the system would stay in equilibrium in its actual state if it were to be submitted not only to the external actions which actually attract them, but also to the fictitious inertial forces. If, therefore, starting from the state in which a system is at a given instant, one imposes upon it a virtual alteration that does not alter the temperature of its various parts, one would impose at the same time upon its internal potential a certain increase, and this increase would equal the sum of the virtual works of the external actions and of the inertial forces.

The equations of motion which flow from this principle are easy to write when the system is represented by a certain number of independent normal variables; they are drawn, in fact, from the equilibrium equations (1) by adding to each of the external actions the corresponding inertial force. If $J_\alpha, J_\beta, \ldots, J_\lambda$, are the inertial forces that refer to the variables $\alpha, \beta, \ldots, \lambda$, these equations are written:

$$A + J_\alpha = \frac{\partial \mathcal{J}}{\partial \alpha}, \quad \ldots\ldots, \quad L + J_\lambda = \frac{\partial \mathcal{J}}{\partial \lambda}. \qquad (2)$$

---

[1] See: Part One, Chapter VII: *d'Alembert's Principle and Lagrange's Dynamics*.

These equations, drawn from d'Alembert's Principle, do not suffice for taking account of the motions that one sees in Nature.

Their insufficiency had already been recognised by the Old Mechanics by analysing certain purely local motions. Thus, the study of Hydrodynamics had shown that such equations could not render an exact account of the motion of fluids; the divergences observed, collected together under the name of *phenomena of viscosity*, had already led Navier to complicate the preceding formulae by the introduction of new terms.

This insufficiency is noted still more exactly in the analysis of certain facts which entirely escaped the grasp of the Old Mechanics.

Amongst the normal variables which, in conjunction with the temperature, define a system, sometimes there is one whose value may change without any of the elementary masses that compose the system being displaced in space; such a variable does not appear in the expression for the kinetic energy, and similarly for the *generalised velocity* which corresponds to it; from then, the method given by Lagrange for calculating the inertial forces shows that the inertial force with respect to this variable is always zero; one has to do with an *inertialess variable*.

Here is an example of an inertialess variable:

In the body of a rigid receptacle there is a homogeneous mixture of chlorine, hydrogen and hydrochloric acid; in order to fix the state of such a system it suffices to add to the temperature a single normal variable, the degree of acidity of the gaseous mixture. When the value of this variable comes to increase, a certain mass of free hydrogen and free chlorine is transformed into an equal mass of hydrochloric acid; but each of the elementary masses which compose the mixture keeps an invariant position in space; this affirmation assumes, clearly, that one does not attribute the act of combination to some hidden motions, to some displacements of atoms, hidden from observation; but it is precisely the character of the New Mechanics to exclude the consideration of such motions from the schemes that it constructs for representing reality. The degree of acidity of the mixture is therefore an inertialess variable.

Amongst the Equations (2) which control the motion of the system, an inertialess variable would given an equation identical with the corresponding equilibrium condition (1). In particular, if the state of the system were to depend upon one single variable apart from the temperature and this variable were to be inertialess, the equilibrium conditions would have to be satisfied at each instant; at each instant the system would be precisely in the state wherein it would remain in equilibrium if the temperature and the external actions would cease to vary. If one were to take to a given temperature a mixture of hydrogen, chlorine and hydrochloric acid, contained in a rigid receptacle, this mixture would immediately exhibit the degree of acidity that would assure equilibrium at the temperature considered.

Experiment shows that this is not so; the composition of such a system varies from one instant to another; equilibrium is only reached at the end of a somewhat long time.

The consideration of inertialess variables thus exploded before our eyes this verity that the analysis of local motions had already discovered: d'Alembert's Principle, accepted without any modification, does not admit the establishment of a General Dynamics.

What alteration is going to be made to this Principle? This alteration is, in some way, imposed by the hypotheses made in the study of viscous fluids since Navier.

To each of the normal variables other than the absolute temperature, one will make correspond not only an external action and an inertial action, but even a *viscous action*; each viscous action will depend not only upon the temperature and other normal variables which determine the state of the system, but even *generalised velocities*, that is to say, upon derivatives with respect to time of the various variables other than the temperature; in addition, these viscous actions will possess an essential property which will permit them to be regarded as retarding actions, as *passive resistances*; *when there is no real modification of the system, they will not perform a positive work*: for certain real motions they will be able to produce a zero work; this will always hold if the system suffers, in space, a displacement of a kind that does alter neither the configuration nor the state; but, in general, the work of the viscosity will be negative.

The state of the system at each instant is no longer the equilibrium state that it would present if it were subjected at the same time to the external actions and to the inertial forces; it is the state in which it would stay in equilibrium if it were submitted simultaneously to the external actions, to the inertial forces and to the viscous actions. Hence, if one imposes upon this system a virtual displacement which does not alter its temperature, these three sorts of actions will perform virtual works and the sum of these three kinds of works will have to be equal to the increase suffered by the internal potential.

Such is the Principle upon which rests the whole of General Dynamics.

Briefly sketched by Helmholtz[1], it has received its explicit enunciation in our own researches[2], and extended by the works of L. Natanson[3].

---

[1] Helmholtz: Ueber die physikalische Bedeutung des Princips der kleinsten Wirkung, *Borchardt's Journal für Mathematik*, vol. C, (1886), pp. 137,213; *Abhandlungen*, vol. III, p. 203.

[2] P. Duhem: Commentaire aux Principes de la Thermodynamique, 3e partie, *Journal de Mathématiques*, 4e série, vol. X, (1894), p. 203; Théorie thermodynamique de la viscosité, du frottement et des faux équilibres chimiques, *Mémoires de la Société des Sciences physiques et naturelles de Bordeaux*, 5e série, vol. II, (1896).

[3] L. Natanson: Various Memoirs, starting in 1896, published in: *Bulletin de l'Académie des Sciences de Cracovie*; and in: *Zeit-*

Let us assume that the system is defined by its temperature and by a certain number of mutually independent normal variables $\alpha,\ldots,\lambda$; let us assume that $v_\alpha,\ldots,v_\lambda$ are the viscous actions which correspond to these variables; the equations of motion of the system will no longer be the equations (2), but the equations:

$$A + J_\alpha + v_\alpha = \frac{\partial \mathcal{J}}{\partial \alpha}, \quad \ldots\ldots, \quad L + J_\lambda + v_\lambda = \frac{\partial \mathcal{J}}{\partial \lambda}. \quad (3)$$

Let us stop for a moment at these equations which, to a great measure, condense the teachings of the New Mechanics.

The external actions and the derivatives of the internal potential introduced into these equations the various normal variables, including the temperature; the viscous actions depend, in addition, upon *generalised velocities*, that is to say, first derivatives with respect to the time, upon normal variables, excluding the temperature; to these various quantities the inertial forces add the generalised accelerations, that is to say the second order derivatives, with respect to time, of the same variables. The equations (3) are therefore, in general, what geometers call *second order differential equations*.

This character of the Equations (3) implies the following consequence:

The motion that a system, subjected to some given actions, takes on starting at a given instant is not determined if one knows only the state of the system at the initial instant; in general, it is necessary to add the knowledge of the initial values taken by the generalised velocities.

But this law, which is the very foundation of Classical Dynamics involves some exceptions in General Dynamics.

When a normal variable is inertialess, the generalised acceleration which corresponds to it disappears from the equations (3). In particular, if the system is defined exclusively by the inertialess variables, the equations of motion cease to be second order differential equations, to be no more than first order equations. From then the motion taken, after starting from a certain instant, by the system subject to some given actions is determined solely by the knowledge of the initial state and without recourse to the initial velocities.

This remark is important. In fact the systems which interest the chemist are, almost always, defined by some variables which correspond to some vanishing or negligible inertial forces. The Dynamics of the inertialess systems therefore implies, to a very large part, Chemical Dynamics. What we have just said suffices to show that several propositions true in the Dynamics of Local Motions will not be able to be extended to the Dyamics of Chemical Reactions; however, these two Dynamics, apparently incompat-

---

schrift für physikalische Chemie.

ible, are drawn from one and the same General Dynamics; but one of them is the more often derived by neglecting the viscous actions, and the other by deleting the inertial forces[1].

---

[1] P. Duhem: Théorie thermodynamique de la viscosité du frottement et des faux équilibres chimiques, *Mémoires de la Société des Sciences physiques et naturelles de Bordeaux, 5e série*, vol. II, (1896); *Traité élémentaire de Mécanique chimique, fondé sur la Thermodynamique*, Vol. I, (Paris), (1897), livre II, p. 201; *Thermodynamique et Chimie; leçons élémentaires à l'usage des chimistes*, (Paris), (1902), p. 455.

CHAPTER X

SUPPLEMENTARY RELATIONS

These reflections are not the only ones that the general equations of the motion give rise to.;
In order that the motion of a system be determined, it is necessary — in the general sense that we give to the word motion — to know at each instant the value of the temperature and of the normal variables; the determination of the motion is therefore the determination, as functions of the time, of the temperature and of the normal variables.

Each of the normal variables, excluding the temperature, furnishes one of the equations (3) which control the motion; it is therefore quite clear that the number of these equations is less by one than the number of functions to be determined[1].

If the system is composed of various parts taken to different temperature, the number of unknown functions will again exceed the number of equations of motaon furnished by Thermodynamics; the excess would be equal to the number of the independent temperatures that there are to consider.

The principles put forward up to now therefore do not suffice for completely putting the general problem of Dynamics into equations; in order that this be put into equations without any omission, it is necessary, in the relations already obtained, to add as many *supplementary relations* as there are distinct temperatures to determine, and these relations have to be drawn from new principles.

What will these principles be?

Let us decompose the system into parts, each of which, at each

---

[1] P. Duhem: *Hydrodynamique, Élasticité, Acoustique, Vol. I,* (Paris, (1891), pp. 18,99; Commentaire aux Principes de la Thermodynamique, 3e partie: Les Équations générales de la Thermodynamique, *Journal de Mathématiques pures et appliquées, 4e série,* vol. X, (1894), chapitre II, p. 225.

instant, will have a uniform temperature, whilst the temperature will not be able to be the same for two different parts. The principles which we have posed suffice to calculate the quantity of heat released, during an infinitesimal time, by each of these parts. This calculation, moreover, brings into evidence a result that we have to indicate incidentally[1].

Let us take the quantity of heat released by each of the parts of the system; let us divide it by the absolute temperature of this part; let us form the sum of the quotients so obtained, and let us add to that the increase suffered by the entropy of the system; the value so obtained is, in general, positive. This proposition states, in the most comprehensive form, the celebrated *Clausius Inequality*[2] which, so powerfully and so happily, has influenced the evolution of Mechanics. However, in certain exceptional cases the sum we have just formed is zero; this takes place, in particular, if all the viscous actions are zero; thus, for the systems without passive resistance that the Old Mechanics studied, the Clausius inequality is transformed into an equality.

But let us come back to the formation of the supplementary relations.

The calculation of the quantity of heat released by each of the parts of the system brings into play the external actions, the internal potential, the inertial forces, that is to say the temperatures, the normal variables, the generalised velocities and the generalised accelerations. It is therefore a function of all these quantities, or of some of them, which will be evaluated as the quantity of heat released by each part of the system.

Let us now assume that the hypotheses, distinct from those that we have invoked up to here, furnish us with another expression for this same quantity of heat; from the comparison between these two expressions there pops up a relation between the variables which fix the state of the system; we shall obtain thus, as well as the supplementary relations, that there are, in the system, parts, or, in other terms, that there are temperatures independent of one another.

This second expression for the quantity of heat that each of the parts of the system yields to the contiguous parts is furnished to us by the theory of heat exchanges that permits conduction. This theory, conceived by Fourier, as is known, thus becomes the indispensible auxiliary of Thermodynamics; it alone renders possible the formation of the supplementary relations, without which the setting in the form of an equation of the problem of Dynamics would be incomplete.

---

[1] P. Duhem: Commentaire aux Principes de la Thermodynamique, 3e partie: Les Équations générales de la Thermodynamique, *Journal de Mathématiques pures et appliquées*, 4e série, vol. X, (1894), pp. 228,238; *Théorie thermodynamique de la viscosité, du frottement et des faux équilibres chimiques*, (Paris), (1896), p. 41

[2] See: Part One, Chapter XII: *The Impossibility of Perpetual Motion*.

The study of the propagation of heat by conduction from one region to the other of the system is linked in an intimate and inextricable way to the study of the motion of this system; any one of these two problems cannot be treated independently of the other. At least, this is so in general. But the dissociation of these two problems, ordinarily impossible, becomes possible in certain particular cases; the cases treated by the Old Mechanics are amongst this number.

From then following question[1] is forced upon our attention: Which are the systems whose motion can be studied without recourse to the supplementary relations? And, immediately, that question is transformed into this one: Which are the systems whose equations of motion, such as Thermodynamics gives, do not contain the temperatures of the various bodies?

Such systems must not be affected by viscosity, for the viscous actions most certainly depend upon temperature; the equations which control their motion are therefore not the Equations (3) but simply the Equations (2). If one investigates which are the systems where the Equations (2) do not contain the temperatures of the various parts, one immediately finds that these systems are characterised in the following way: Their internal potential is the sum of two terms; the first term depends upon the temperatures of the various parts and not upon the other normal variables; the second term does not depend upon temperatures and depends only upon the other variables.

Here there are some very remarkable systems; in the course of thermodynamical deductions they are met at every instant apart from exceptional cases. One essential property flows from the form of their internal potential: under a real or virtual modification which leaves unchanged the temperature of each part, these systems release none and absorb no heat; for these, every *isothermal* modification is, at the same time, an *adiabatic* modification.

One can easily give an example of such *isothermal-adiabatic systems*: it is sufficient to take a collection of bodies each of which keeps an unchanged shape and to assume that the state of each of these bodies is entirely defined by its position in space and by the distribution that the temperature imposes upon it. Now, such a collection well represents the general type of systems that the Old Mechanics studied. One thus understands that one can determine the motion of such systems without making any appeal to the theory of conduction; that, for them, the establishing of the equations of Dynamics preceded the discovery of the laws of propagation of heat. The formulae which control this

---

[1] P. Duhem: Sur l'équation des forces vives en Thermodynamique et les relations de la Thermodynamique avec la Mécanique classique, *Procès verbaux de la Société des Sciences physiques et naturelles de Bordeaux*, (Meeting of 23rd December, 1897); L'intégrale des forces vives en Thermodynamique, *Journal de Mathématiques pures et appliquées*, 5e série, vol. IV, (1898), p. 5.

propagation only come into play, once the motion of the system is known, for studying the variations of the temperature of the different bodies; once the motion of the stars is determined by Celestial Mechanics, one can, with Fourier, propose to determine the distribution of the temperatures on each of them.

This resolution *in two steps* of the problem of Dynamics is possible, as we have said, only for *isothermal-adiabatic systems*; the motion of no other system can be determined if one does not take account of the supplementary relations. Geometers were compelled to recognise this truth as soon as they wanted to treat the propagation of sound in air, to analyse a system foreign to this very particular category; the correction applied by Laplace to the expression for the speed of sound that Newton gave consisted essentially in substituting a supplementary relation for another.

CHAPTER XI

THE KINETIC ENERGY EQUATION

AND USABLE ENERGY

The kinetic energy equation played an essential rôle in the in the development of the Old Mechanics[1]; let us investigate what it becomes in the New Mechanics[2]; this question is going to lead us to the consideration of the form taken by the supplementary relations.

Under every virtual modification without an alteration in temperature, the sum of the works of the external actions, of the inertial forces and viscous actions is equal to the increase of the internal potential.

Let us write the equality which expresses this proposition upon attributing as a virtual variation, for each of our normal variables, precisely the real variation that it suffers, in an infinitesimal time, by the effect of the motion of the system. The virtual work of the external actions, the virtual work of the inertial actions become the real values that these actions perform in the time considered; the virtual work of the inertial forces become the decrease that the kinetic energy of the system suffers in the same time; as for the increase that the internal potential would suffer in a virtual isothermal modification; it would not become equal to the increase that the same quantity would suffer in the real modification, for, ordinarily, the latter would no longer be isothermal.

Therefore, in general, the excess of the work that the exter-

---

[1] See: Part One, Chapter VII: *d'Alembert's Principle and Lagrange's Dynamics*.

[2] P. Duhem: Sur l'équation des forces vives en Thermodynamique et les relations de la Thermodynamique avec la Mécanique classique, *Procès verbaux de la Société des Sciences Physiques et Naturelles de Bordeaux*, (Meeting of 23rd December, 1897); L'Intégrale des forces vives en Thermodynamique, *Journal de Mathématiques pures et appliquées*, 5e série, vol. IV, (1898), p. 5.

nal actions and the inertial actions actually perform during a certain lapse of time over the increase that the kinetic energy suffers during the same lapse of time, can only be equal to the increase that a certain quantity, entirely determined by the state of the system, would take.

But this proposition, in general false, can be true in certain particular cases, and these cases it is essential to know. Let us therefore investigate the circumstances in which the second member of the *Kinetic Energy Equation*, the translation of the preceding proposition, becomes the increase in a certain quantity which depends only on the state of the system. When such a quantity exists we shall name it the *usable energy* of the system; the reason for this we shall now see.

First of all, can a system admit a usable energy whatever may be the form attributed to the supplementary relations?

In order for this to be so, as is seen without difficulty, the internal potential must be the sum of two terms, of which one depends exclusively upon the temperatures and upon no other normal variables, whilst the other depends upon the normal variables and upon none of the temperatures; in other words, the system considered must be transformed into an *isothermal-adiabatic* system if one deprives it of its viscous actions; furthermore, in such a system the usable energy is identical with the internal potential, which does not depend upon temperatures.

Amongst the systems that we have just defined are those that the Old Mechanics studies.

The other systems would not admit a usable energy in all circumstances and for whatever form the supplementary relations may take. But certain particular forms attributed to these relations may assure them of such an energy. It is this that happens, in particular, when the supplementary relations make invariable the temperature of each of the parts of the system, when, consequently, all the real modifications are *isothermal*; these are precisely the conditions which are fulfilled in a perfectly conducting system the surface of which is maintained at a uniform and unchanging temperature. The usable energy is then identical with the internal potential.

There is another case where the system admits a usable energy by virtue of the supplementary relations; this is the case where these relations transform the entropy of each of the parts of the system into a function of the temperature of this part alone; moreover, the quantity which plays the rôle of the usable energy changes with the form of this function.

This case is realised in its simplest form when the entropy of each of the parts of the system necessarily keeps an unchanged value under every real modification. For a system all the real modifications of which are *isentropic*, the usable energy is identical with the internal energy.

This case is not devoid of all physical meaning.

If the system is exempt from viscosity and if the absence of conductibility prevents its different parts from exchanging any quantity of heat between each other or with external bodies, in the course of the motion each of its parts keeps an unchanged entropy. One meets, in Physics, some systems which are approximately subject to such conditions; the motions of a gaseous mass whose conductibility and viscosity are very weak are essentially isentropic motions; in fact this was admitted by Laplace touching upon the motions that propagate sound in air, whilst Newton assumed them to be isothermal.

After having enumerated the various cases where a system admits a usable energy, it remains for us to justify this terminology.

When one assembles some bodies and requires them to be subject to some modifications which, taken together, make a motor, one can propose to produce two kinds of effects from them. One can ask them to displace or to modify certain parts of the system contrary to the tendencies of the external actions, or, in other terms, to oblige the external actions to perform a negative work; of a crane or of a lift one iwll demand a heavy load to be lifted. One can also ask them to increase the kinetic energy of a part of a system; one uses a bow or a cannon to hurl a projectile.

It is therefore natural to take as the measure of the *useful mechanical effect* of a modification carried out in a system the increase of the kinetic energy of the system less the work of the external actions.

If it is a question of a system which admits a usable energy, we shall immediately draw from the kinetic energy equation the following proposition: The useful mechanical effect exceeds the decrease in the usable energy by a quantity equal to the work of the viscous actions. Now one recalls that the real work of the viscous actions can never be positive. The preceding proposition can therefore be stated in the following way: The useful mechanical effect of a modification can never exceed the loss of usable energy that the system suffers in this modification; in general it is inferior to it; exceptionally, it is equal to it if the modification implies no work by the viscous actions.

This proposition justifies the terminology usable energy.

If all the modifications of the system are isothermal, the rôle of usable energy is taken, as we have said, by the internal potential; whence the terms *available energy* and *free energy* that Gibbs, Maxwell and Helmholtz attributed to this potential. But the internal potential only keeps this rôle for isothermal modifications; for isentropic modifications, for example, it is yielded to the internal energy; whence the importance of this latter to examine the useful effect of a charge of powder which explodes in a container impermeable to heat and the name of *explosive potential* that it takes in ballistics.

CHAPTER XII

STABILITY AND DISPLACEMENT

FROM EQUILIBRIUM

The notion of usable energy gains its greates importance in discussions about stability and a state of equilibrium. When a system admist a usable energy, when, in addition, the virtual work of the external actions is the decrease of a potential, entirely determined by the state of the system, the celebrated proposition of Lagrange[1], the rigorous proof of Lejeune-Dirichlet, all apply themselves; the equilibrium assuredly is stable in a state where the sum of the usable energy and of the external potential has a minimum value.

If it is a question of one of these exceptional systems for which there exists a usable energy whatever the form given to the supplementary relation, no restriction will either complicate the statement or limit the scope of the preceding theorem; this is so in the domain of the Old Mechanics as well.

This is no longer so if the system only admits a usable energy by virtue of the particular form attributed to the supplementary relations; in this case the minimum of which the preceding proposition speaks no longer has to be such that the usable energy increases with every virtual modification starting from the state which corresponds to this minimum, but only in every virtual modification where the form of the supplementary modifications is preserved; in addition, stability would not be assured by the criterion that we have just announced if the real motions of the system did not preserve the same supplementary relations; it goes without saying, moreover, that the usable energy in question is that which flows from the particular form attributed to the supplementary relations.

Let us assume, for example, that the sum of the internal energy and the external energy has a minimum value, not amongst all

---

[1] See: Part One, Chapter VII: *d'Alembert's Principle and Lagrange's Dynamics*.

the values that this sum can take, but amongst all those that it can take without any of the system's parts changing entropy; the system is certainly in stable equilibrium if it is only capable of isentropic modifications; but the stability is no longer guaranteed if the system can assume motions which are not isentropic, for example isothermal motions; *isentropic stability* of the equilibrium does not imply *isothermal stability*. If one wishes to guarantee the isothermal stability of the equilibrium, one does not have to make a minimum of the sum of the internal energy and external potential, but the *total potential*, that is to say the sum of the internal potential and external potential; and this total potential must not be rendered a minimum for every virtual modification imposed upon the system, but for every virtual modification which does not alter the temperature.

A system, as we have said, that is placed in a certain state of equilibrium would perhaps lose all stability if it ceased to have isothermal motions. On the contrary, in an equilibrium state where isothermal stability is assured, the isentropic stability also is[1].

The proof of this proposition necessitates that one make appeal to a hypothesis which must be regarded as one of the fundamental principles of Thermodynamics; we have proposed that this hypothesis should be named *Helmholtz's Principle*, for Helmholtz stated it explicitly[2] without, however, regarding it as a distinct principle.

Let us imagine that the state of a system of which all the points are at the same temperature be defined by the absolute value of this temperature and by a certain number of other normal variables; let us imagine also that in keeping its value at each of the latter we gave the absolute temperature an infinitesimal increase; the system absorbs an infinitesimal quantity of heat; the ratio of the quantity of heat absorbed in the increase of the temperature is a quantity whose value depends only upon the state of the system; it is the *normal calorific capacity* of this system.

The calorific capacity of every system is positive; such is Helmholtz's Postulate.

This postulate, free of its algebraic form, takes a very simple and very sharp concrete meaning. Plainly, it can be stated thus: To raise the temperature of a system, to warm it up without causing it to suffer any change of state, it is necessary to provide it with heat, it is necessary to *heat* it. Put in this form,

---

[1] P. Duhem: Commentaire aux Principes de la Thermodynamique, 3e partie: Les équations générales de la Thermodynamique, *Journal de Mathématiques pures et appliquées, 4e série*, vol. X, (1894), chapitre IV, p. 262; *Traité élémentaire de Mécanique chimique fondée sur la Thermodynamique, Vol. I*, (Paris), (1897), livre I, chapitre X, p. 163.

[2] Helmholtz: Zur Thermodynamik chemischer Vorgänge, I, *Sitzungsberichte der Berliner Akademie*, (1er Semester, 1882), pp. 12,19; *Abhandlungen*, vol. II, pp. 969,978.

Helmholtz's Postulate appears as the justification of the words *quantity of heat* used to designate an algebraic symbol which appeared without any link with the notion of temperature, starting from our sensations of hot and cold.

But it is not necessary to be mistaken about the scope of the new statement and to believe that it confers upon Helmholtz's Postulate some experimental evidence; it contains some obscure and ambiguous phraseology; to raise the temperature of a system *without making it suffer any other change of state* is an expression whose meaning changes with the nature of the variables that are associated with the temperature in order to determine the state of the system.

Whilst true when these variables are normal variables, the proposition would no longer be true in other cases. In fact, the study of vaporisation of liquids has introduced the use of certain non-normal variables and the consideration of a certain specific heat relative to these variables, the *specific heat of the saturated vapour*; now, in certain circumstances the specific heat of the saturated vapour may be negative.

The case where the Helmholtz Postulate is definitely true is distinguished from other cases in a very precise character; in the first case, an alteration of temperature without alteration of state does not imply work by external actions; this is not so in the other cases; one can therefore make precise, in the following way, the concrete statement of this postulate: To raise the temperature of a system without producing either an alteration of state or *external work*, it is necessary to provide it with heat; it is necessary to remove heat in order to lower this temperature.

It is thanks to Helmholtz's Postulate that in a system which suffers no alteration of state, which gives rise to no external work, and which is enclosed in a container of uniform and constant temperature, conduction and radiation tend to make the temperature everywhere equal to that of the container; through this consequence Helmholtz's Postulate is connected with the ideas of Sadi Carnot and Clausius.

Instead of warming up a system defined by normal variables, without making it suffer any other alteration of state, one can heat it up by keeping unchanged the external actions that it experiences; one is then led to consider the *calorific capacity for constant actions*; if the conditions of isothermal stability are fulfilled, the calorific capacity for constant actions is greater than the normal calorific capacity; it is therefore positive. For example, the specific heat of a gas at constant pressure is larger than the specific heat at constant density, as was also recognised by Laplace and Poisson.

The study of isentropic stability, of its relations with isothermal stability, still leads to some interesting remarks; to review them would take up too much space, so let us pass them by in order to halt at the principal consequences of the criterion of isothermal stability.

The New Mechanics extends to new domains the application of

the proposition of Lagrange and Lejeune-Dirichlet, and this extension is immense.

The Old Mechanics could legitimately draw from this proposition the conditions which suffice for guaranteeing the stable equilibrium of an incompressible liquid or again that of a solid floating on the surface of such a liquid. In the case where the external forces reduce to weight, the first problem offers no difficulty; the second has been solved by Bravais and Guyon. But the study of compressible fluids exceeded the scope of the classical methods. The New Mechanics, on the contrary, can give the conditions which suffice for guaranteeing the isothermal stability of equilibrium for a compressible fluid whose elements do not act upon each other, that this fluid exists alone or that it bears a single float[1].

The various problems that are raised by the study of electricity and magnetism also offer numerous opportunities for applying the new methods; let us cite some of them.

Can a mass of soft iron placed in a magnetic field and deprived of all support stay in equilibrium? According to an old legend, the coffin of Mahomet remained thus, floating in air, in a Mosque in Medina. If the distribution of magnetism over the mass of soft iron is stable, kept immobile, the equilibrium of this mass becomes strongly unstable when there is restored to it the faculty of moving in every direction[2]; the least breath of air would suffice to precipitate Mahomet's coffin to the ground or to one of the magnets that attracted it.

Faraday explained the phenomena presented by diamagnetic bodies, such as bismuth, by supposing that these bodies had a negative coefficient of magnetisation. Now, on such bodies the magnetic distribution would not possess isothermal stability. This result, obtained by Beltrami and us[3] at the conclusion of simultaneous

---

[1] P. Duhem: *Hydrodynamique, Élasticité, Acoustique, Vol. I*, (Paris), (1891), livre II, chapitre II, p. 80; Sur la stabilité de l'équilibre des corps flottants, *Journal de Mathématiques pures et appliquées, 5e série*, vol. I, (1895), p. 91; Sur la stabilité d'un navire qui porte du lest liquide, *ibid.*, vol. II, (1896), p. 23; Sur la stabilité de l'équilibre d'un corps flottant à la surface d'un liquide compressible, *ibid.*, vol. III, (1897), p. 151.

[2] P. Duhem: *Théorie nouvelle de l'aimantation par influence fondée sur la Thermodynamique*, (Paris), (1888), chapitre IV, §2; *Lecons sur l'Électricité et le Magnétisme, Vol. II*, (Paris), (1892), p. 215.

[3] E. Beltrami: Note fisico-matematiche, *Rendiconti del Circolo matematico di Palermo*, vol. III, (Meeting of 10th March, 1889); P. Duhem: *Comptes rendus de l'Académie des Sciences*, vol. CV, (1887), p. 798; *ibid.*, vol. CVI, (1888), p. 736; *ibid.*, vol. CVIII, (20th May, 1889), p. 1042; Des corps diamagnétiques, *Travaux et Mémoires des Facultés de Lille*, (1889); *Lecons sur l'Électricité et le Magnétisme, Vol. II*, (Paris), (1892), p. 221.

and independent investigations implies the rejection of the hypothesis formulated by Faraday; almost forcibly it leads to accepting another hypothesis that Edmond Becquerel had given out; the aether in the vacuum would be magnetic and the diamagnetic bodies would simply be bodies less magnetic than the aether; from there one can, by way of analogy, draw a most valuable argument in favour of the electrical theories of Maxwell and Helmholtz which attribute to the aether a dielectric property.

It is in the domain of Chemical Mechanics, if not very accessible to the theories of the Old Mechanics, that the new methods and, in particular, the theory of isothermal stability give their most fecund consequences. The study of the stability of equilibrium appears as the indispensible complement of Statics; touching this study, Gibbs gave a few indications; today it has taken some great strides.

Here is the point of departure for these developments:

Several fluids are mixtures of fluids, but incapable of exercising any chemical reaction upon each other; their elements do not act upon each other; they are screened from every external action, save that of a uniform and constant normal pressure. Homogeneity evidently characterises the state of equilibrium of such a mixture. Does this equilibrium state possess isothermal stability? This certainly is possessed, provided there hold certain conditions which are easily found.

Having obtained these conditions, let us admit that they are satisfied by the various mixtures which form a chemical system; in other words, let us admit that each of these mixtures would be capable of having an equilibrium state endowed with isothermal stability if one were to deprive of chemical activity the various bodies which comprise it. If we then give to these bodies the faculty of giving rise to reactions, we shall be able to state[1] the following propositions:

If the chemical system is maintained at a constant pressure, or if it is enclosed in a container of constant volume, in neither case can it present unstable isothermal equilibrium; all the chemical equilibria reached under these conditions are stable or indifferent.

The isothermal indifference at constant pressure characterises the equilibrium states of a whole category of chemical systems, of those where the number of *phases* surpasses the number of *independent components*. Isothermal stability, on the contrary, is the general rule when the number of phases is at most equal to the number of independent components.

Nevertheless, if isothermal stability is then the rule, this rule has exceptions. The systems where the number of independent

---

[1] P. Duhem: On the General Problem of the Chemical Statics, Journal of Physical Chemistry, vol. II, (1898), pp. 1,19; Traité élémentaire de Mécanique chimique, fondée sur la Thermodynamique, Vol. IV, (Paris), (1899), livre, IX, chapitres I,III, pp. 281, 346.

components is equal to the number of phases are able to present states of equilibrium which are unaffected if one maintains unchanged the temperature and pressure; and these *unaffected states*, which property makes them exceptional, excite a high degree of interest in physicists; some theorems of great importance, discovered by Gibbs and rediscovered by Konoralov, characterise them.

Also, the simplest of these unaffected states have been first indicated by some experimenters, Bakhuis Roozeboom and Guthrie; they are produced when a hydrate of a certain substance is in the presence of an aqueous solution of this substance and that the solution has exactly the same compositions as the hydrate. Some more complicated unaffected states have been either observed by chemists or predicted by theoreticians, notably by Paul Saurel[1].

The systems whose equilibrium states possess isothermal stability are subject to laws which provide the experimenter with extremely valuable qualitative indications; we shall want to speak of laws about the *displacement of equilibrium*.

Amongst the various laws that can be brought under this title we shall choose two, which in Chemical Mechanics are of a considerable importance: The law of *isothermal displacement under variation of pressure* and the law of *displacement under variation of temperature*. These two laws were formulated in 1884, the latter by J.H. van't Hoff, and the former by H. Le Chatelier; but it is the Mechanics, alone, based upon Thermodynamics that has permitted us to rigourously link them to the notion of stability and even to enunciate them in a completely correct way[2].

A chemical system is in equilibrium and this equilibrium would be stable if one kept the temperature and pressure constant.

Without changing the temperature one gives the pressure an infinitesimal increase, then one fixes its new value. The equilibrium of the system is disturbed; after a certain lapse of time a new state of equilibrium is established, infinitesimally near to the first; of what nature is the chemical reaction that the system must suffer in order to pass from the first equilibrium state to the second? The answer to this question is very simple. The reaction in question must present the following character: Performed at the original pressure and the original temperature, both kept constant, it would make the volume of the system decrease.

---

[1] Paul Saurel: *Sur l'équilibre des systèmes chimiques*, (Bordeaux Thesis), (1900).

[2] P. Duhem: Sur le déplacement de l'équilibre, *Annales de la Faculte des Sciences de Toulouse*, vol. IV, N, (1890); Sur le déplacement de l'équilibre, *Annales de l'École Normale Supérieure, 3e série*, vol. IX, (1892), p. 375; Commentaire aux Principes de la Thermodynamique, 3e partie: Les équations générales de la Thermodynamique, *Journal de Mathématiques pures et appliquées, 4e série*, vol. X, (1894), p. 262; *Traité élémentaire de Mécanique chimique fondée sur la Thermodynamique, Vol. I*, (Paris), (1897), livre I, chapitres VIII-XI.

Thus the decomposition of hydrochloric gas by oxygen gives water vapour and chlorine; carried out at a constant temperature and constant pressure, this reaction is accompanied by a decrease in volume. At a specified temperature and a specified pressure there is established between the four gases a stable state of equilibrium; an increase of pressure without a variation in temperature determines the production of another state of stable equilibrium; one must pass from the first state to the second by decomposing a certain quantity of hydrochloric gas by oxygen. Hence, at a given temperature, the mixture will be the richer in chlorine as the pressure is the higher.

This example shows at the same time how important and how easy to deduce are the propositions about displacement of equilibrium by the variation of pressure. The law of displacement of equilibrium under variation imply consequences important in another way and throws another kind of light upon the field of Chemical Statics.

A system may be in equilibrium at a specified temperature and, for example, at a specified pressure. If one keeps the temperature and pressure unchanged, this equilibrium would be stable. Without changing the pressure, one give the temperature an infinitesimal increase; the system suffers a transformation which leads it to a new state of equilibrium. What do we know about this transformation? This much: if carried out at the initial pressure and the initial temperature, it will absorb heat.

A mixture of oxygen, hydrogen and water vapour, subject to a specified pressure, such as atmospheric pressure, is in stable equilibrium at a specified temperature; the composition of this mixture is also specified. Let us raise the temperature a little whilst keeping the mixture at atmospheric pressure; the composition of the mixture in equilibrium is going to change; the reaction which produces this alteration will have to absorb heat if it is carried out at constant temperature and constant pressure; hence it consists in the destruction of a certain quantity of water vapour, since the water vapour is an exothermal compound. Thus for an unchanged pressure the water vapour suffers dissociation that is the more complete the higher the temperature.

The principle of the displacement of equilibrium under variation of temperature should show us, similarly, that an endothermic compound heated at constant pressure is decomposed the less the higher the temperature.

These propositions solve one of the most important and the most debated problems in Chemistry.

In fact, since the end of the eighteenth century chemists have tried, in a different way from the sign of the quantity of heat brought into play, to compare *exothermic reactions* and *endothermic reactions* with each other.

First they thought that every combination was exothermic, that every decomposition was endothermic. By discovering endothermic compounds, P.A. Favre gave the *coup de grâce* to this theory.

Then the Thermochemical Hypothesis, formulated by Kelvin, saw in exothermic compounds those which can be directly formed from their elements, and in endothermic compounds those which can be decomposed spontaneously. Experiment has done justice to this long triumphant hypothesis.

It is here that General Mechanics, based upon Thermodynamics, presents us in a new light the comparison between exothermic combinations and endothermic combinations, whilst a raising of the temperature determines a destruction of the first, it favours the formation of the second.

This new comparison, which gives such a good account of dissociations and syntheses produced at very high temperatures, by H. Sainte-Claire Deville, by Debray, and by Troost and Hautefeuille, was noticed by Lavoisier and Laplace[1] and was almost immediately forgotten. The progress of Thermodynamics looks back a quarter of a century to that day; they discovered it first of all for systems in unaffected equilibrium; implied in the geometric constructions of Gibbs[2], but without any statement bringing it to the attention of chemists, its was formulated by J. Moutier[3] in a Note of some pages length which, with a stroke of genius, broke the clouds over Thermochemistry; J.H. van't Hoff[4] next extended it to chemical equilibria which are stable.

Thus, the laws which control the displacement of equilibrium have changed the face of Chemical Mechanics; their scope, furthermore, does not stop within the confines of this domain, however vast; if time did not prevent us from following them through transformations and the consequences, we would see them render signal service in the study of elasticity, electricity and magnetism; but we have said enough to allow the reader to judge the fertility of the theorem of Lagrange and Lejeune-Dirichlet extended by the New Mechanics.

---

[1] Lavoisier et de Laplace: Mémoire sur la chaleur, read to the Académie des Sciences, 18th June, 1783, *Mémoires de l'Académie des Sciences pour l'année 1780*, pp. 387-388.

[2] J. Willard Gibbs: On the Equilibrium of Heterogeneous Substances, *Transactions of the Academy of Connecticut*, vol. III, (January, 1876), p. 181; *Équilibre des systèmes chimiques*, (Paris), (1899), p. 110.

[3] J. Moutier: Sur les transformations non réversibles, *Bulletin de la Société Philomathique, 3e série*, vol. I, (1877), p. 39.

[4] J. van't Hoff: *Études de Dynamique chimique*, (Amsterdam), (1884), p. 161.

CHAPTER XIII

FRICTION AND FALSE CHEMICAL

EQUILIBRIA

The Principle of the Conservation of Energy is a strong tree stump whose roots plunge deep into the heart of our first mechanical conceptions; out of this stump grow several trunks; we have just described the principal one: the Statics formulated by Gibbs is at its root; it is continued by the Dynamics which Helmholtz has given us a sketch of.

This trunk is the first which was to shoot from the Principle of the Conservation of Energy; it is therefore the most developed and vigorous; it is not alone; beside it others have grown for a while, which, for a moment, must hold our attention.

In fact, Gibbs' Statics and Helmholtz's Dynamics, however vast they may be, do not suffice for embracing the immensity of physical phenomena; there are some modifications which do not submit to their laws, some systems which can not be represented by their formulae.

The systems which follow the rules of this Statics and this Dynamics are defined with precision by a certain character: Every continuous sequence of equilibrium states of such a system is a reversible modification. If a system does not present this character, it cannot be submitted to the rules of this Statics and of this Dynamics; the means of putting it into equilibrium and the laws of its motion have to be demanded from another Statics and another Dynamics.

Now, we have already met some systems whose equilibrium states, arranged in a continuous sequence, do not form a reversible modification[1]; these are the systems susceptible to permanent alterations; it is in that that we are warned first of all that there is a place for creating a special Statics, a new special Dynamics for the systems which are able to suffer permanent alterations.

These are not the only systems which beg for the creation of

---

[1] See: Part Two, Chapter VI: *The Reversible Modification*.

a particular Mechanics; we are going to define another category whose requirements will not be less.

What is a reversible modification? It is a continuous sequence of states of equilibrium; but, in addition, it is the frontier common to two groups of real modifications directed in two senses opposite to each other. Let us suppose that a reversible modification links the two extreme states A and Ω. It will be possible to determine an infinitely slow real modification taking the system from A to Ω; this modification makes it pass through a sequence of states of which each differs infinitesimally from one of the states which form the reversible modification; furthermore, in these two infinitesimally close states the system is subject to infinitesimally close external actions. It will also be possible to determine a real modification taking the system from Ω to A, and endowed with analogous properties.

Let us now imagine that in studying a physical system we ascertain the following particularity: In general, given an equilibrium state, if one modifies it infinitesimally, *and in an arbitrary manner*, the properties that the system possesses in this state, the external actions that affect it, one is led to a new state of equilibrium. It is clear that a continuous sequence of such states of equilibrium cannot be a reversible modification; for another sequence of states, infinitesimally close to the first, will again be a sequence of equilibrium states; this will not be able to be a real modification. A system which offers a similar particularity therefore does not present the character by which are recognised the systems subject to the Mechanics of Gibbs and Helmholtz; it requires the creation of another Mechanics.

Let us make precise the character which marks this new category of material systems. This character is the following: For each of these systems one can conceive of states of equilibrium such that in every state sufficiently close to one of them the system stays in equilibrium if one subjects it to actions sufficiently close to those which maintain it in equilibrium in the first state.

Examples of similar systems abound: let us borrow the first from Chemical Mechanics.

At elevated temperatures, at 1,500°C or 2,000°C, a mixture of oxygen, hydrogen and water vapour presents the token by which one recognises systems subject to the Mechanics of Gibbs and of Helmholtz; at a given temperature and a given pressure, the mixture in equilibrium has a specified composition; if one changes but a little this composition without changing either the temperature or the pressure, one breaks the equilibrium of the system; increasing just a little the proportion of water vapour, one creates a mixture at the heart of which the water vapour dissociates; upon diminishing this same proportion, one creates a mixture at the heart of which the oxygen and hydrogen combine; coordinating the temperature and pressure according to whatever law one may want, one obtains a continuous sequence of states of equilibrium and *this sequence is a reversible modification*.

FRICTION AND FALSE CHEMICAL EQUILIBRIA

Everything is quite different at low temperatures, at 100°C or 200°C; here, whatever may be the composition of the mixture, whatever may be its content of water vapour, chemical equilibrium is assured; there is produced neither dissociation nor combination. Therefore let us take such a mixture at 200°C and at atmospheric pressure; let us assign it successively, in thought, all possible compositions, starting from that which corresponds to the total absence of water vapour up to that obtained by pushing to the limit the combination of oxygen and hydrogen; we obtain a continuous sequence of states of equilibrium, but not a reversible modification; for, starting from an arbitrary one of the states that compose the sequence we shall be able to alter by small arbitrary amounts the pressure, the temperature, and the composition without the system ceasing to be in equilibrium.

The study of purely local motion, the object of the Old Mechanics, gives rise to some analogous considerations.

On a surface that has a highest point and slopes away from this point on all sides, let us place a very small weighted body whose contact with the surface is not free from friction. It is not only at the highest point that this small body will stay in equilibrium; it is also true on the gradients provided that they are not too steep; it will also be possible, around the summit, to delimit a certain area around every point in which the small heavy body will remain immobile; an arbitrary line traced in this area will define a continuous sequence of equilibrium states, but not a reversible modification; for, starting from an arbitrary one of these equilibrium states one will be able to disturb the mobile point, to change a little the force which acts upon it; always will it remain at rest.

The mechanical system which provides us with this very simple example is also going to provide the name by which we shall designate the category of material systems that occupies us at this moment; we shall call them *frictional systems*.

It is therefore the Static and Dynamics of Frictional Systems, essentially distinct from the Statics and Dynamics developed up to here, that must now be treated.

But at the very threshold of this investigation one objection delays us: Do there actually exist frictional systems? Are not the particularities that we have thought to observe and which have served us to define them simple illusions? Do they not vanish when they are subjected to a minutely detailed analysis?

According to the majority of mechanicians, a solid body which slides or rolls on another does not rub; but a multitude of small roughnesses bristle upon the two surfaces in contact; they foul upon each other, interlock, hook, and break; and the friction is only a fiction by which one encompasses these innumerable and complicated, imperceptible phenomena without analysing them.

At 100°C and 200°C a mixture of oxygen, hydrogen and water vapour seems in equilibrium whatever may be its composition; according to several physicists this equilibrium is only apparent; in reality the oxygen and hydrogen combine, but extremely slowly, with such a slowness that observation in laboratories can divulge no trace of this combination; this slowness alone differentiates the phenomena observed at low temperatures and phenomena observed at high temperatures.

What exactly is the scope of these objections?

There is no doubt that two rough surfaces chafe more energetically upon each other than two smooth surfaces; one would be able to conclude that two bodies touching by strictly smooth surfaces will not chafe; the averred existence of a fictitious friction, synthesising roughness and deformations, does not suffice to exclude the possibility of a real friction. Besides, Hydrodynamics[1] obliges us to consider frictions other than the mutual friction of two solid bodies; it shows us that a liquid chafes upon a solid, that two superposed liquids chafe upon each other along the length of their common surface; what unevennesses, what roughnesses, what hidden gearings will be invoked, in this latter case, to reduce friction to an appearance?

On the other hand, the reality of an equilibrium state can always be denied; where one thinks to see a system in equilibrium, another can, without fear of contradiction, pretend that there is a motion, but a motion so slow that the most prolonged observations do not allow one to ascertain any alteration in the system. Pushing this plea to the bar to the extreme, J.H. van't Hoff did not hesitate[2] to regard the time that has elapsed since the period of colliery to our time as too short for certain chemical systems to have suffered an appreciable transformation. But this opinion cannot be used to control experiment; if experiment is incapable of contradicting it, it is not less incapable of confirming it; it would be necessary, for it to be able to provide an evidence, that it be extended to durations compared with which the geological periods are only a moment; further, if its evidence were unfavourable, could one always challenge it and require tests that lasted for a still longer time.

Such a loophole evidently has only one aim: To subject the whole of Physics to the laws of Statics and Dynamics which were formulated by Gibbs and Helmholtz. It would have a logical value if we could recognise from another source the legitimacy of this goal, if we had reasons to believe that all mass systems have to yield to the rules of this Statics and this Dynamics. But of such reasons we have none. To define the systems that are able to yield to these rules, we have, amongst all conceivable systems, cut out a certain group; we have made this excision in an arbitrary way by this hypothesis posed *a priori*: Every continuous sequence of states of equilibrium of one of the systems considered forms a reversible modification.

---

[1] P. Duhem: Recherches sur l'Hydrodynamique; 4e partie: Les conditions aux limites, *Annales de la Faculté des Sciences de Toulouse*, 2e série, vol. V, (1903).

[2] J.H. van't Hoff: *Archives néerlandaises des Sciences exactes et naturelles*, série II, vol. VI, (1902).

Experiment proved that our hypothesis was useful, that it was not a vain whim without a real object; that the line of demarcation it traced—and which could only enclose the smallest parcel of land—delimits a vast and fertile domain. The Mechanics of systems of reversible modifications is shown to be apt for representing, to a sufficient approximation, a great number of physical phenomena. Does that authorise us to think that all the phenomena produced in inanimate Nature have to be set in order by this Mechanics? Our hypothesis was only, in the proper sense of the word, a *definition*; in the immensity of the possible it circumscribes an infinitely particular case. From the fact that this particular case represents a good part of the real world, are we correct in concluding that it comprehends the whole of reality? Do we have to, at any price, enclose the entirety of Physical Nature in this little islet around which there extends infinitely the ocean of systems that reason can conceive? Is it permitted to us, in this goal, to reject the most obvious evidences, the most sure, the best controlled of experiments, in favour of unverifiable affirmations? Is it not more logical to think that that which appears to our senses as a particular case is also, in Nature, only a particular case? That outside of systems whose states of equilibrium can always be arranged in reversible modifications there exists an infinity of other systems whose Statics is not the Statics of Gibbs, whose Dynamics is not the Dynamics of Helmholtz, and that amongst these systems there are ranged exactly the systems endowed with friction.

Hence the laws according to which frictional systems move or keep in equilibrium claim a particular formula. This formula will not be called for haphazardly. The formula imposed upon Statics by Gibbs and upon Dynamics by Helmholtz is shown to be admirably fertile; it is natural to safeguard the type as much as possible; to draw out the new formula from the old formula by means of additions as modifications as light as possible; it is this idea which served us as a guide when we constructed the Mechanics of frictional systems[1].

It would be inconvenient to expound the latter here without entering into details that this work is not concerned with; let us try, nevertheless, to outline a brief sketch and, to this end, let us restrict ourselves to the study of a system that one single normal variable, besides the temperature, suffices to define.

Let us represent this unique variable by the letter α; if $\mathcal{F}$, A, J, v are the internal potential, the external action, the inertial force and the viscous action, we can, according to the Dynamics of Helmholtz[2], write at each instant the equality:

---

[1] P. Duhem: Théorie thermodynamique de la viscosité, du frottement et des faux équilibres chimiques, *Mémoires de la Société des Sciences physiques et naturelles de Bordeaux*, 5e série, vol. II, (1896); Recherches sur l'Hydrodynamique, 4e partie, *Annales de la Faculté des Sciences de Toulouse*, 2e série, vol. V, (1903).

[2] See: Part Two; Chapter IX: *The Principle of General Dynamics*.

$$A + J + v = \frac{\partial \mathcal{F}}{\partial \alpha}. \tag{3}$$

This equality, the general law of motion of the system, implies the laws of its equilibria, a law conforming with Gibbs' Statics.

The equilibrium of frictional systems does not conform with Gibbs' Statics; the equality (3) is therefore no longer applicable to them; but one can attempt to modify it in such a way that it does extend to such systems.

To this end one will continue to attach to each state of the system a quantity $\mathcal{F}$, specified unambiguously by the knowledge of this state; one will continue to attach to this quantity, which will still be called the internal potential, the internal energy and the entropy by the relations known beforehand; the external action, the inertial force, the viscous action, will remain as before; but these elements will no longer suffice for posing the equation of motion of the system; it will be necessary to know a new element, the *frictional action f*.

This action, always positive, will depend, like the viscous action, upon the absolute temperature, upon the variable α, upon the general velocity $\alpha' = d\alpha/dt$; but contrary to what holds for the generalised velocity, it will also depend upon the external action A; in addition it will not vanish at the same time as the generalised velocity; as the latter tends to zero the frictional action will tend towards a positive value g.

To control the motion of the system we shall no longer have a single equation, but two distinct equations; the first will only have to be employed if the generalised velocity $\alpha' = d\alpha/dt$ is positive; it will have the following form:

$$A + J + v - f = \frac{\partial \mathcal{F}}{\partial \alpha}. \tag{4}$$

The second will be written:

$$A + J + v + f = \frac{\partial \mathcal{F}}{\partial \alpha}. \tag{4'}$$

It will be reserved for the case where the generalised velocity $\alpha' = d\alpha/dt$ is negative.

As for the equilibrium condition, it will no longer be represented by an equality but by a double inequality expressing that the absolute value of the difference $A - \partial\mathcal{F}/\partial\alpha$ does not exceed g:

$$-g \leqslant A - \frac{\partial \mathcal{F}}{\partial \alpha} \leqslant g. \tag{5}$$

Let us pass rapidly over what touches upon the kinetic energy equation; one can repeat here almost what has been said in studying Helmholtz's Dynamics; one has only to add to the work of the viscosity the work of the friction, and the latter, like the former, is always negative. Let us also pass over the Clausius inequality which stays in tact in the New Dynamics; there, again, the work of the friction is only added to the work of the viscosity. Other consequences and laws which have just been formulated, and particularly the equilibrium condition, are going to hold us up a little longer.

The Statics of Gibbs would require that the difference $A - \partial \mathcal{F}/\partial \alpha$ be zero, and therefore lie between $-g$ and $+g$, the states of equilibrium that this Statics predicts, and which are usually called the states of *true equilibrium*, are thus the same in number as those the new Statics predicts; but the latter declares the existence of an infinity of other states of equilibrium, which are designated by the name of states of *false equilibrium*.

If the value of g is large, the states of false equilibrium spread out, on either side of the states of true equilibrium, into a vast domain; they shrink, on the contrary, close to the true states of equilibrium if the value of g is small; if this value becomes sufficiently slight the states of false equilibrium should diverge so little from the states of true equilibrium that experiment would no longer be able to distinguish between them; in practice the Statics of Frictional Systems would then be identical with Gibbs' Statics.

Here there is only one particular application of the following remark: Gibbs' Statics, Helmholtz's Dynamics are limiting forms of the Statics and Dynamics of frictional systems; the latter tend to the former when the frictional actions become infinitesimal.

This remark is not a simple view of the mind; it takes a particular interest in the study of chemical equilibria[1].

The better to fix our attention, let us choose an example studied with great care by Ditte and Pélabon. In a sealed tube let us heat some selenium liquid that is surmounted by a mixture of the vapours of selenium, hydrogen, and hydroselenic acid. Whilst the temperature does not exceed 150°C the system keeps in equilibrium, whatever the composition of the gaseous mixture; when the temperature is raised any more the system becomes capable of chemical reaction; if the gaseous mixture is weak in hydroselenic acid, the selenium and hydrogen combine; if the gaseous mixutre is rich in hydroselenic acid, this compound dissociates. At a given temperature one observes the phenomenon of *com-*

---

[1] We have expounded the theory of chemical equilibria taking account of friction and the principal applications of this theory in the following writings: *Théorie thermodynamique de la viscosité, du frottement et des faux équilibres chimiques*, (Paris), (1896); *Traité élémentaire de Mécanique chimique fondée sur la Thermodynamique*, Vol. I, (Paris), (1897), Livre II; *Thermodynamique et Chimie, leçons élémentaires à l'usage des chimistes*, (Paris), (1902), leçons XVIII,XIX,XX.

bination whilst the acidity of the mixture is less than a certain limit; one observes, against this, the phenomenon of *dissociation* whenever the acidity exceeds another limit, and this second limit is higher than the first; when the acidity lies between these two limits the mixture is certainly in equilibrium.

In proportion as the temperature is raised, the acidity which limits from above the zone of combination and the acidity which limits from below the zone of dissociation approach each other; the zone of equilibrium grows thinner; at 325°C its width becomes indiscernible; the theoretician may well still assume that there subsists a frictional action very small in value; but for the experimentalist it no longer produces states of true equilibrium subject to Gibbs' Statics.

What we have just observed in this example is a particular case of a general rule; in every chemical system frictional actions become weaker when the temperature is raised; very large and low temperatures hinder every chemical reaction; starting from a certain temperature, which varies with the chemical system studied, reaction becomes possible, but it is limited by false equilibria; then, when the temperature attains a sufficiently raised degree the region of false equilibria becomes so narrow that the experimentalist can no longer discern it; in practice one no longer observes only true equilibria tracing the common boundary between two reactions going in opposite directions; a sequence of such states of equilbrium forms a reversible modification.

It is therefore only when the temperature exceeds a certain limit, varying from one chemical system to another, that one can use laws of the Statics announced by Gibbs and his followers; never would one have been able to extend to chemical reactions the laws of this Statics if one had been restricted to considering transformations produced at low temperatures; this extension would have been impossible if H. Sainte-Claire Deville had not had the genius to demand of the chemistry of very high temperatures the secret of Chemical Mechanics. The service which, by that, he rendered Science is comparable to that which Galileo rendered to the study of local motion when, making the abstraction of friction, he dared to announce the law of inertia.

It was necessary, in order that Science might being to be developed, that this Statics of Gibbs, which is a very simplified Statics, might be expounded first of all; but, because this Statics is a very simplified Statics, the development of Chemical Mechanics would soon be held up if one did not seek how to complete it; in particular, chemical reactions taking place at room temperature, the ones that are produced all the time in our laboratories, stay incomprehensible. The introduction of friction unravelled this chaos; by the consideration of false equilibria the influence of the temperature upon chemical transformations ceases to be a mystery; the study of the stability of these same equilibria is the key to the theory of explosions.

Furthermore, the phenomena of false equilibria is not met only in the study of purely chemical equilibria; the vaporisation of certain solids is sometimes arrested by similar equilibria, and they will probably be met in the study of the freezing of liquids[1]. Thus is affirmed the universal necessity of a Mechanics from which frictional actions are not banished.

---

[1] P. Duhem: Sur la fusion et la cristallisation et sur la théorie de M. Tammann, *Archives néerlandaises des Sciences exactes et naturelles, 2e série*, vol. VI, (1901), p. 93.

CHAPTER XIV

PERMANENT ALTERATIONS

AND HYSTERESIS

When one arranges in a continuous sequence a set of states of equilibrium presented by a frictional system, one does not obtain a reversible modification; by this character frictional systems escaped the grasp of the most usual theorems of Thermodynamics; they necessitate a special Statics and a special Dynamics. The impossibility of forming a reversible modification by arranging a set of states of equilibrium into a continuous sequence is not the exclusive property of frictional systems; we have also met it in studying a system capable of permanent deformations[1].

The systems having permanent alterations therefore approach frictional systems because the notion of reversible modification is inapplicable to the one just as to the other; but there the analogy stops. Some essential differences separate these two categories of systems.

Let us consider an equilibrium state of a frictional system; in general it is not possible to take a system to this state by a modification which is always infinitely slow, no more than so to lead if from this state; certain single exceptional states of equilibrium can be met by a modification of extreme slowness.

Let us take, to the contrary, an equilibrium state of a system capable of permanent alteration; a modification of a slowness that is always infinitesimal can lead a system to such a condition, it can also lead it away from such a condition. But let us imagine that, in order to draw the system of this state by an infinitely slow modification, we have varied the temperature and the external actions according to certain laws; let us cause this temperature and these external actions to pass through the same set of values, but in reverse order; the system will suffer a new infinitely small modification which will not be the simple reversal of the first, which will not make it go back through the same states,

---

[1] See: Part Two; Chapter VI: *The Reversible Modification*.

which, generally, will not lead it to the initial equilibrium.

The theory of systems capable of permanent alterations will therefore be distinct from the General Mechanics whose principles, after Gibbs and Helmholtz, we have sketched; but it will also be distinct from the Mechanics of frictional systems; this will be a new branch of Mechanics.

How is this New Mechanics going to be constituted?

The principal idea alone interests us here; the detail of formulae would not be in place in this work; let us therefore limit ourselves to the study of a simple case which will the better make clear the outlines of the idea; let us choose as the object of our analysis a system defined by a single normal variable outside the temperature; for example, a tensioned wire for which the length will be this normal variable, whilst the tensioning weight will be the corresponding external action.

First of all let us give to the temperature and to the tensioning weight certain infinitesimal variations; the length of the wire suffers an infinitesimal increase. Let us next give to the temperature and to the tensioning weight some variations equal in absolute value to the preceding ones, but opposite in sign, of such a kind that there two quantities recover their initial value; the length of the wire decreases; but this decrease does not have the same absolute value that the preceding increase suffered, for the wire remains affected by a permanent deformation.

Thus in the course of an infinitely slow modification a linear algebraic relation determines the infinitesimal variation that the length of the wire suffers when one is given the infinitesimal variations imposed upon the temperature and the tensioning weight; but this relation does not have to have the same form when the wire is lengthened and when it is contracted; a certain equality has to be written when the normal variable suffers a positive variation, and another when it suffers a negative variation.

What guide will help us to discover the form of these two equalities? The same theory, which cannot suffice for treating permanent alterations, but which is shown to be so fertile in the study of systems with reversible modifications. We shall seek to construct our New Mechanics in such a way that it approaches the former theory as much as possible, that it flows from it by a very simple transformation, that it is a generalisation of it, that the Statics and the Dynamics of systems of permanent alterations can be regarded as limit forms of the Statics and the Dynamics of systems having very small permanent alterations. In a word, we shall follow a method similar to that which gave us the theory of frictional systems.

When a system exempt from permanent alteration suffers an infinitely slow modification, that is to say a reversible modification, the equilibrium conditions are, at each instant, satisfied; if the state of the system depends upon a single normal variable α, the external action A is equal, at each instant, to the derivative of the internal potential $\mathcal{F}$ with respect to α, $\partial \mathcal{F}/\partial \alpha$; it is this that teaches us[1] the Equations (1).

Amongst infinitesimal variations, coordinated mutually, of the temperature, the external action and the normal variable, there then exists the relation:

$$dA = d\frac{\partial \mathcal{F}}{\partial \alpha}, \qquad (6)$$

by virtue of which the quantities always equal to A and $\partial F/\partial \alpha$ simultaneously suffer equal increases. According to this relation, if one changes the signs of the variations which the temperature and the external action suffer without changing their absolute value, one changes the sign of the variation suffered by the normal variable without any longer changing the absolute value; by that is expressed the reversibility of the infinitely slow modification.

These particularities would not be met in a system capable of permanent alterations; each of the elements whose succession forms an infinitely slow modification can no longer be controlled by the equality (6); for this equality we have to substitute two distinct relations, the one valid when the normal variable increases only, the other valid when this variable decreases only.

In the first case we shall substitute for the equality (6) the relation:

$$dA = d\frac{\partial \mathcal{F}}{\partial \alpha} + hd\alpha; \qquad (7)$$

in the second we shall substitute for it the relation:

$$dA = d\frac{\partial \mathcal{F}}{\partial \alpha} - hd\alpha. \qquad (7')$$

The quantity h, whose introduction in these equations distinguishes the systems capable of permanent alterations from those which are exempt from them, depends upon the state of the system, and also upon the external action A.

It suffices, which goes without saying, to give to this quantity h a value sufficiently small for the equalities (7) and (7') to differ very little from the equality (6); the permanent alterations of the system are then very sensitive and its infinitely small modifications are nearly reversible; thus the systems exempt from permanent alterations and capable of reversible modifications are presented to us as limiting forms of systems affected by small permanent alterations.

For the systems exempt from permanent alterations a very simple rule permits us to draw from the internal potential the knowledge of the internal energy and, therefore, of the quantity of heat brought into play in an infinitely slow modification. Noth-

---

[1] See: Part Two, Chapter VIII: *Internal Potential and General Statics*.

ing hinders one from extending this rule to systems capable of permanent alterations. Coupled with what has preceded, it will furnish the essential principles upon which reposes the Statics[1] of such systems. Some additional hypotheses, all inspired by the desire to render the new bough of Thermodynamics as similar as possible to the main branch, will complete these principles.

What are the applications of the New Statics?

A first category of permanent alterations is formed from elastic deformations. Traction, torsion, bending, involve deformations which do not disappear with the causes that produce them; these deformations, known and observed from antiquity find their theoretical explanation in the preceding principles.

The remanent magnetism that a piece of steel keeps after the magnetising action has finished has to be set amongst the number of the most remarkable of permanent alterations; despite the investigations of G. Wiedemann, which had already brough into evidence the direct relations between the residual elastic deformations and the remanent magnetism, the laws of this latter phenomenon were kept singularly obscure; they have been clarified in these latter years, principally by the investigations of Ewing and his disciples; Ewing has given the name of *magnetic hysteresis* (ὑστέρησις, delay) to the property that iron has of retaining the remanent magnetism. The ideas introduced by Ewing in the study of magnetic hysteresis have infiltrated little by little into the analysis of other permanent alterations and have rendered this analysis more fertile; also the same word hysteresis is commonly adopted today to designate the aptitude of an arbitrary system to permanent alterations.

The polarisation of dielectric bodies offers such analogies with the magnetisation of magnetic bodies that one must, beside magnetic hysteresis, place dielectric hysteresis, although the latter is, up to now, very much less known than the former.

Essential in the study of elasticity and in the theory of magnetism, hysteresis appears to be called to play a very important rôle in Chemical Mechanics. The investigations of experimentalists multiply from day to day the cases where one observes permanent alterations of the physical state of the chemical constitution; amongst these investigations let us particularly cite the minute determinations of Gernez on the various transformations of sulphur, the patient experiments of van Bemmelen on the absorp-

---

[1] We have devoted six Mémoires to the exposition of this Statics, collected under the title: Les déformations permanentes et l'hystérèsis, *Mémoires in-4° de l'Académie de Belgique,* vol. LIV, (1895); *ibid.,* vol. LVI, (1898); *ibid.,* vol. LXII, (1902); eight Memoires published under the title: Die dauernden Aenderungen die Thermodynamik, *Zeitschrift für physikalische Chemie,* vol.s XXII,XXIII, (1897); *ibid.,* vol.'s XXVIII,XXXIII, (1899); *ibid.,* vol. XXXIV, (1900); *ibid.,* vol. XXXVII, (1901); a paper entitled: On the Emission and Absorption of Water Vapor by Colloidal Matter, *Journal of Physical Chemisty,* vol. IV, (1900), and various other papers.

tion of water vapour by gelatinous silicon and by other jellies.

It is, without doubt, to permanent alterations of this kind that one has to refer the effects of tempering, annealing, and hammer-hardening, which so strangely complicate the study of metals and of their industrial combinations. Very often these effects result at the same time as elastic hysteresis and chemical hysteresis; alone, the simultaneous consideration of these two hystereses unravelled some little bit the phenomena, apparently inextricable, that certain bodies present; such are the nickel steels whose strange properties Ch.-Ed. Guillaume has analysed, or the silver-platinum alloy whose electrical resistance manifests, according to H. Chevallier, such curious residual variations.

This superposition of chemical hysteresis upon elastic hysteresis makes the laws of expansion of glass singularly complex; the observation of the displacements that the zero point of thermometers suffers had only just revealed, first to Despretz and next to Ch.-Edmond Guillaume, something other than this extreme complexity; by numerous and patient observations, guided by the Thermodynamics of permanent alterations, have at last allowed L. Marchis to put some order into this chaos.

We shall not, as is clear, be able to show here how the Statics of which we have just outlined a first sketch is applied to very complex and varied phenomena; we shall restrict ourselves to indicating, rather, how to analyse certain essential ideas which are discovered in the course of this development.

In a system affected by permanent alterations, the quantity h, which we shall henceforward name the *hysteresis coefficient*, is non-zero in general; the two equalities (7) and (7') are therefore distinct from each other; if we assume that the system suffers, with an infinite slowness, an infinitesimal modification due to certain variations of the temperature and of the external action, upon reversing these variations we shall not be able to reverse the modification and lead the system to the initial state.

But that which is not true in general may become so in certain particular cases; upon suitably associating the values of the normal variable, of the temperature, and of the external action, one can annul the hysteresis coefficient; when these values are associated in such a way, we shall say that the system is placed in a *natural state*; in general, if one takes the system in an arbitrary state, defined by a certain value of the normal variable and a certain value of the temperature, one will be able to submit it to an external action such that this state becomes natural.

For infinitesimal modifications issuing from a natural state the two equalities (7) and (7') are identical with each other and with the equation (6); in other terms, every infinitesimal and infinitely slow modification issuing from a natural state is a reversible modification; if one imposes upon the temperature and the external action some small variations followed by equal and opposite variations, one takes the system back exactly to its first state; it keeps no permanent alteration.

It is quite different when the state is not a natural state.

Let us impress upon the values of the temperature and the external action a small oscillation, which deflects them a little bit from what they first were and then takes them back; the system preserves a permanent alteration that notes a change of value of the normal variable. This residual alteration, it is true, is very small; but, if the temperature and the external action suffer in their values a new oscillation, a new residue will be added to the first. Thus, in impressing upon the temperature and the external action some very small and very numerous deflections, now in one direction and now in another, followed by returns to values which always stay the same, we shall see the system suffer a gradual and considerable alteration owing to the accumulation of very small, but very numerous, residual alterations.

One successively sees the scope of this remark.

In the world there are not unchanging temperatures, nor unvarying actions; the most perfect processes of adjustment narrow the limits between which the values of these elements oscillate; they are not suppressed by the variations. These incessant, inevitable, but imperceptible, variations of the external actions and of the temperature generate, in the long run, a noticeable alteration of the state of the system; this state therefore seems to change spontaneously whilst the conditions in which the system is placed seem unchanged.

Instead of reducing to the extreme, by artifices of adjustment, the incessant perturbations that the external actions suffer, one can exaggerate them by a systematic disordering; then, apparently spontaneously, one also exaggerates the alterations which the bodies actually experience. Thus there is explained the influence, so often experienced, of shocks impressed upon a tensioned or twisted thread; of shocks, of vibrations, of alternating currents applied to a magnet; of diurnal variations of the temperature modifying the glass of a thermometer. The experimental investigations of Ewing, of Tomlinson, of van Bemmelen, of L. Marchis, of Lenoble, of H. Chevallier, abound in interesting remarks about the rôle of *shocks*.

The apparently spontaneous modifications that a system suffers when the temperature and the external action stay *practically* unchanged allow one to characterise this system and to set it in one or the other of the two categories we are going to define.

In the first category the change generated by the accumulation of very numerous and very small residual alterations ceaselessly draws the state of the system near to the natural state which fits the almost unchanged values of the temperature and of the external action; when the system has arrived at this natural state, the very small and incessant alterations of the conditions in which it is placed no longer cause it to suffer any appreciable modification. For such a system the natural state feigns being mistaken for the stable equilibrium state of a system devoid of hysteresis; it possesses almost all its properties. The experimental investigations of Ewing bring into evidence, with great nicety, this character presented by the natural state of a magnet.

Under the influence of small perturbations that the temperature and external action suffer, the systems of the second category shun the natural state that the values of this temperature and this action characterise. The consideration of the circumstances under which a wire tensioned by a weight ceases to belong to the first category and passes to the second, clarifies the phenomena, such as *elongation with constriction*, which precede rupture.

This rapid sketch lets us catch a glimpse of the extent and variety of questions to which is applied the Statics of systems affected by hysteresis; it is clear, however, that this Statics would not exhaust the study of systems such as this; it teaches us of the properties which infinitely slow modifications enjoy; but an infinitely slow modification is only the limit of a real modification; every real modification is pursued with a finite speed and its study requires the formation of a Dynamics.

For systems devoid of hysteresis, the passage from the Statics to the Dynamics was assured, first of all, by d'Alembert's Principle; for the external action it sufficed, according to this principle, to substitute the sum of the external action and of the inertial action. The extension of this principle to systems affected by hysteresis had to be presented in the spirit of a simple and most natural hypothesis; in fact, the formulae drawn from this hypothesis allow the analysis of some phenomena ascertained by experimentalists.

However, long investigations were not needed to establish the insufficiency of this Dynamics founded upon the extension of d'Alembert's Principle; plainly it did not take account of the particularities that systems affected by hysteresis present when they are rapidly modified; the observations of Bouasse and Lenoble on deformations of wires by torsion or by traction, the investigations of Max Wien and other observers on the magnetisation of iron in an alternating field have brought into evidence some particularities which escape the grasp of this Dynamics.

Indeed, one should not be astonished at that. The Dynamics based upon d'Alembert's Principle is constantly in default in that same study of systems devoid of hysteresis; to render it acceptable, it has been necessary to complicate it, to add to the external action not only inertial action, but yet a viscous action. Is it not quite natural that the analysis of systems affected by hysteresis causes to pop up in front of our eyes the same insufficiency of d'Alembert's Principle, the same necessity of appealing to a more complicated hysteresis? Is it not also quite natural to make this hypothesis follow that which is shown to be fertile in the theoretical examination of systems without permanent alterations, to pass again from the Statics to the Dynamics by substituting for the external action the sum of this action, of the inertial force, and of a viscous action?

By this assumption the Dynamics of systems affected by hysteresis is created[1]; in agreement with observations, too few up to now, which permanent deformations of systems in motion have been subjected to, it waits for new stimuli to its development from experiment and for new opportunities to be subjected to the inspection of facts.

Furthermore, this Dynamics of systems affected by hysteresis resembles, through an essential character, the Dynamics of systems with reversible modifications and the Dynamics of frictional systems; in every closed cycle traced by an arbitrary one of these systems Clausius' inequality is satisfied; after a sequence of modifications none of these systems can revert to its primitive state without having produced an essentially positive uncompensated modification; this constant sense in which all modifications of the Universe are oriented is imposed with the same rigour upon all motions.

---

[1] contra: P. Duhem: Les déformations permanentes et l'hystérèsis, VII, Hystérèsis et viscosité, *Mémoires in-4° de l'Académie de Belgique*, vol. LXII, (1902).

CHAPTER XV

ELECTRODYNAMICS

AND ELECTROMAGNETISM

Beside the principal trunk of Thermodynamics, beside the Mechanics of systems without friction or hysteresis, we have seen rise up two other stems, still young and the development of which is a very long way from being accomplished: the Mechanics of frictional systems and the Mechanics of systems with hysteresis. These two stems are not distinguished, first of all, from the principal trunk; up to a certain height they remain knit together, identical with it; they separate from it only at the moment where the Mechanics of systems devoid of friction and hysteresis invokes the notion of reversible modification. All that which precedes the use of this notion, all that appeals to the Principle of the Conservation of Energy alone is common to the three Mechanics.

Issued from the same roots, a fourth stem arises, born a long time ago and robust already; it treats the Mechanics of electric currents; but, with the first three stems this off-shoot has only the stump in common; it does not knit together with them; from what has been said up to now touching upon the Conservation of Energy, almost nothing may be applied straight away to Electodynamics and to Electromagnetism.

We have constantly admitted, in that which precedes, that the properties of a system at a given instant were to be characterised entirely by two kinds of elements; in the first place, the values of a certain number of variables which define the *state* of this system; in the second place, the speeds of the various material points in the local motion which activates the system. The Total Energy of the system depends upon these two kinds of elements; the first alone appear in the expression for the Internal Energy; by means of the second ones, one forms the live force or the Kinetic Energy. If the position of some part of the system depends upon some one of the independent variables, the total energy of the system depends not only on this variable, but, further, upon its derivative with respect to the time or, according to the denomination that we have adopted, of the corresponding *generalised velocity*; this latter appears in the expression for the live force. Contrary to this, if the value of a certain independent value does not influence the position of the various parts of the system, the generalised velocity which corresponds to this inertialess variable only comes into play in the formula which determines the internal energy of the system.

These principles are at the very root of the various branches of Mechanics of which we have, up to here, followed the development; they become false for the systems that electric currents cover.

The properties that are possessed, at a given instant, by bodies through which electric currents run do not depend alone — as has been known since Ampère — upon the way in which the electricity is distributed at this instant; to fix these properties it does not suffice to say what is the electric density at each point of a conducting mass or upon a surface bounding such a mass; it is necessary to say further what are, at each point of the conductor, the components of the *electric flux*: now to give these components is to give the derivative with respect to time of every electric density, the generalised velocity which corresponds to a similar density. Thus, even if the electric density is an inertialess variable, the generalised velocity which corresponds to it influences the actual properties of the system; the latter do not depend solely upon the *state* of the system, not solely upon its local motion; they depend, in addition, upon the *electric motion* which is the root of it; one must predict, from now, that an alteration of electric motion will correspond to a certain work, that the energy of the system will depend upon this motion, that besides the internal energy and the kinetic energy, it will include an *electrokinetic energy*.

More nicely still are these ideas manifested in the study of polarised dielectrics; the properties of such a dielectric, at a given instant, are not entirely fixed when one knows, at this instant, the magnitude and direction of the *polarisation intensity* at each point of the medium. Since Maxwell and, above all, since Helmholtz, no one doubts whether it is necessary to join together the magnitude and direction of the displacement flux; now the components of this flux are simply the generalised velocities which correspond to the components of the polarisation. Here again, the properties of the system are only completely determined if one knows the generalised velocities corresponding to certain inertialess variables; one must expect the introduction of the generalised velocities in the formula which expresses the total energy of the system.

It is therefore a new Mechanics, distinct from that we have expounded up to here, that will raise up the study of systems through which electric currents run; if we refuse to recognise this point, if we were to try to construct an Electrodynamics which flows from the principles previously adopted, the most flagrant disagreements will spring up between theory and experiment.

If we were to form the energy of an electrified system by introducing only the values taken at each instant by the electric density and the polarisation, without taking account of the generalised velocities with respect to these inertialess variables, that is to say the conduction flux and displacement flux, we would be able, by the principles that we have posed, to construct an Electrical Statics which would fully agree with the facts; to pass from this Statics to Electrical Dynamics it would suffice to know the laws which are obeyed by the viscous actions in an electrified system; some very simple hypotheses, accepted since Ohm, furnished us with these laws.

The equations of motion of electricity that we would then be led to write would not be without use; they would agree with those that Kirchoff gave for metallic conductors at a uniform temperature, that Kelvin formed for thermoelectric cables, that Gibbs and Helmholtz applied to electrolytes. But exactly at the times when the electrical motion would be reduced to a permanent regime of immobile conductors, these equations would fall in default or when the currents would vary whenever the conductors are set in motion; then there would be produced effects of *electrodynamical induction* that they could not predict.

We would also be able to draw from the principles with which we are familiar the forces which tend to displace or to deform the various parts of the system; the forces calculated thus would not coincide with real forces; amongst them we would not see how to calculate the electrodynamic forces of which Ampère determined the laws.

The calculation of the heat released in a modification, based upon the rules of General Thermodynamics, would give occasion for the same remarks as for electromotive actions. As long as permanent currents run through immobile conductors, this calculation would provide exact results; these results would be those that Joule and Peltier have observed in studying conductors at uniform temperature, that Kelvin discovered in treating bodies heated unequally, that Helmholtz obtained in developing the theory of electrolysis. But every variation of the currents, every movement of the conductors would give rise to some thermal phenomena not predicted by this calculation.

Electrodynamical forces, electromotive action of induction, release heat within the heart of mobile systems traversed by variable currents, such are the effects that a new branch of Mechanics must analyse.

A set of simply hypotheses, sharpened by some appeals to experiment, furnishes the expression for the electrokinetic term that has to appear in the Total Energy[1]. Once the Electrokinetic

---

[1] The order of exposition of Electrodynamics which is indicated here differs a little from that which we have followed in Volume III of our *Leçons sur l'Électricité et le Magnétisme*, (Paris),

Energy is known, it suffices to postulate that the entropy of the system contains no electrokinetic term, just as it contains no kinetic term; to admit that the viscous actions are, in all circumstances, determined by Ohm's formulae, in order to be in full possession of the Principles of Electrodynamics. From these principles, all the principles that constitute this science, all the laws that govern the electromotive forces of induction, electrodynamics actions, the release of heat in the body that currents traverse, are drawn by regular processes.

The various formulae whose set comprise this Electrodynamics all depend upon the consdieration of a certain quantity, which can be calculated when one knows the form of the various bodies of the system and the distribution of the conduction or displacement current of which they are the stage of action. This quantity introduced into Physics by F.E. Neumann, rediscovered in another form by W. Weber, generalised by Helmholtz, is the *Electrodynamic Potential*. Under a real or virtual modification where each conductor is displaced when engagin the electric fluxes that they traverse, the work of the electrodynamic forces is precisely equal to the decrease in this Potential.

Now, the Electrokinetic Energy is precisely equal to this Potential *changed in sign*; this proposition is assuredly worthy of note, for it makes the Electrodynamic Potential play a rôle quite distinct from that of the potential of the electrostatic forces; this latter appears *with its sign* in the expression for the Total Kinetic Energy of the system; thus is nicely noted, from the use of the Principle of the Conservation of Energy, a profound distinction between the Mechanics of Electrodynamic Actions and General Mechanics.

This essential distinction is not, furthermore, going even to exclude certain comparisons, such as this one, amongst others, which is due to Maxwell: In a system traversed by linear and uniform currents one can, from the Kinetic Energy, draw the electrodynamic forces and eletromotive forces of induction by formulae quite similar to what, since Lagrange, serve to calculate the inertial forces when one knows the expression for the live force. This comparison renders more striking the analogy, already within grasp from what has gone before, between the Kinetic Energy and the Electrokinetic Energy; it would not be necessary, however, to exaggerate its scope; its generality knows no limits, for it does not extend to systems traversed by non-uniform currents. Maxwell saw in that a proof that the electric current is reducible to local motion[1]; for us, it particularly expresses this fact, that the Electrokinetic Energy is homogeneous and of the second degree

---

(1892); the former seemed more natural and more rigorous than the latter; this new order will presently be detailed in a special paper; in it there will be given the mathematical deductions that cannot be found a place for here.

[1] See: Part One, Chapter XI: *The Mechanical Theories of Electricity*.

in the intensities of the currents, like the live force is homogeneous and of second degree in the generalised velocities.

The presence of magnets in a system traversed by currents gives rise to the appearance of electromagnetic effects. One could be tempted to link Electromagnetism and Electrodynamics by taking as fundamental hypothesis the analogy between magnets and the currents that Ampère discovered; each magnetic element would under all circumstances, be exactly equivalent to a small closed current suitably chosen. This method was followed by Maxwell; it provided exact expressions for the forces which are exercised between the currents and the magnets and for the electromotive forces of electromagnetic induction; but it does not always suffice for determining the laws of magnetisation of soft iron by currents, and the information that it furnishes touching the exchanges of heat which accompany this magnetisation are contrary to the facts. Certainly the expression for the Energy of the system is not that which such a method gives.

The Electromagnetic Mechanics can be constructed on the same plan as the Electrodynamic Mechanics and is seated upon the same foundations. The Total Energy of the system will be obtained by taking the Total Energy of the system assumed current-less, and simply adding to it the Electrokinetic Energy, of which the expression is known to us henceforth. The entropy will again be the same as if the system gave no passage to a current, and the viscous actions will always conform to Ohm's formulae. From there, there will be deduced the laws of electromagnetic induction, forces which are exercised between currents and magnets, of magnetisation by currents, and finally the quantity of heat brought into play in any electromagnetic effect; and all these laws will fully agree with the results of experiment.

The formulae thus obtained all depend upon an Electromagnetic Potential; a real or virtual displacement, where magnets involve with their magnetisation, where the electric flux remains constantly linked to conductors, gives rise to a work by forces which are exercised between currents and magnets; this work is the decrease of the Electromagnetic Potential. But, as is well worth noting, this Electromagnetic Potential does not in any way appear in the expression for the Total Energy, since, by hypothesis, the latter does not contain an electromagnetic term. Helmholtz had already been led to this somewhat astonishing proposition by a quite different route; he had drawn it from the comparison of electrodynamic systems with monocyclic mechanisms ; before long it was put into a better light by Vaschy[2] and by ourselves[3]; it is

---

[1] Helmholtz: Ueber die physikalische Bedeutung des Princips der kleinsten Wirkung, *Borchardt's Journal*, vol. CX, (1886), p. 155; *Abhandlungen*, vol. III, p. 224.

[2] Vaschy: *Traité d'Électricité et de Magnétisme, Vol. I*, (Paris), (1890), p. 318.

[3] P. Duhem: *Lecons sur l'Électricité et le Magnétisme, Vol. III*, (Paris), (1892), p. 386.

one of those which the better marks the singular character of Electrodynamic and Electrodynamic Dynamics.

Starting from principles, of which we have given a summary description, the Mechanics is developed with as much logic as fullness; Helmholtz[1] in some imperishable Mémoires[1], unfolded from them the marvelous interlinking, leaving his successors hardly any need to verify any links; in this admirable theory, impeccable deductions link to the first hypotheses all Maxwell's audacious inferences that were so fertilely discovered; at the extremity of boughs sprouted by this new branch of Mechanics, there flourishes the most brilliant flower that the genius of that Scots physicist produced, the *Electromagnetic Theory of Light*[2].

And, nevertheless, we are witnesses to a strange phenomenon which will stupefy the historians of Science in the future. This method, so rigorously logical, which allies, without losing the least shred of truth, the most audacious predictions of modern Electrodynamics with the most solid conquests of Coulomb and Poisson, does not find favour with physicists. Several of the latter seem fixed by a kind of hatred, on the contrary, of the old electrical theories whose fertility is manifested, in the domain of thought as in the domain of action, by some unprecedented discoveries. Ungrateful sons, who strike the breast from which they suckled, they have broken with scientific tradition; at the risk of ruining the surest bases of our knowledge that touches upon electricity and magnetism, they wish only to call for Maxwell; they prefer his inexplicable inconsistencies[4] to the major works of a Gauss or an Ampère; they think that the exactitude of an equation no longer needs to be proved when this equation is in his writings: *Ipse dixit*.

If the new Mechanics were not to oppose with all its forces a similar tendency, it would cease to merit the title that it bears, proudly and legitimately, the Old Mechanics; it would no longer be *Rational Mechanics*.

---

[1] Helmholtz: Ueber die Bewegungsgleichungen der Elektricität für ruhende leitende Körper, *Borchardt's Journal*, vol. LXXII, p. 57; *Abhandlungen*, vol. I, p. 545; Die electrodynamischen Kräften bewegten Leitern, *Borchardt's Journal*, vol. LXXVIII, (1874), p. 273; *Abhandlungen*, vol. I, p. 702.

[2] P. Duhem: Sur la théorie électrodynamique de Helmholtz et la théorie électromagnetique de la lumière, *Archives néerlandaises des Sciences exactes et naturelles, série II*, vol. V, (1901), p. 227.

[3] P. Duhem: *Les théories électriques de J. Clerk Maxwell*, (Paris), (1902); Notes sur quelques points des théories électriques et magnétiques, *Mémoires de la Société des Sciences physiques et naturelles de Bordeaux*, 6e serie, vol. II, (1902).

CONCLUSION

Four parts of Mechanics, distinct from each other, have successively attracted our attention; the systems capable of reversible modifications, the frictional systems, the systems with hysteresis, and lastly the systems traversed by currents, have been able to be studied theoretically, under the condition of invoking hypoteses proper to each of the four categories of adopting formulae whose type waried from one to another.

Does the whole domain of Mechanics reduce to the study of the four categories of systems that we have just enumerated? Next to the four trunks whose growth and blossoming we have followed, shall we not, some day, see a new shoot arise. All that it is permitted to affirm is that no logical reason allows the Mechanics already sketched to be regarded as being the only possible Mechanics. In particular, the study of various radiations which, for some years now, have lavished upon experimentalists opportunities for discoveries, have revealed to them some effects so strange, so difficult to subject to the laws of our Thermodynamics, that one would not be surprised to see a new branch of Mechanics swell up from this study.

Whatever may be the number of doctrines, albeit distinct from one another, into which the New Mechanics is subdivided, is not this multiplicity of doctrines a blemish, a mark of inferiority upon the reputation of the Old Mechanics, so perfectly one. To profess such an opinion it would be necessary to ignore the very relations which unite between them the various branches of Thermodynamics.

When a physicist proposes to construct a mathematical system capable of depicting, with some approximation, a fragment, as small as it may be, of the real world, he soon recognises that everywhere there are produced frictions, permanent alterations, electric currents; he therefore cannot formulate the mathematical theory of an arbitrary set of bodies without taking account at the same time of all these categories of phenomena.

But the complication of some such theory can only be extreme; to wish to construct it all at once would surpass the powers of the human mind; a physicist must then, to start with some chance of success the solution of the problem which is presented to him, first simplify the enunciation of this problem. He commences by taking away frictions, permanent alterations, electric currents, and by studying what remains after all these subtractions. He knows that it outlines only a very summary respresentation of reality, that he will have to touch up and complete the results of this first analysis; but he also understands that this oversimplified analysis is necessary in order for him next to be able to make an attempt at a more detailed theory.

When he has constructed this first theory, which will be like the support of its subsequent constructions, he will return to one after the other of each of the complications that he first had neglected; he investigates what modification his first representation must bear if one wishes that it give the image either of the effects of friction alone, or of permanent alterations, or of phenomena produced by currents. Finally, after these partial and successive essays, he is in a position to return to the various parts of his work, to rearrange them, to weld them together, to make a single doctrine of them whose chapters are linked together logically.

In one of the red chalk drawings of which the Louvre is so proud, follow the work of successive approximations by which Raphael created one of the personages that he painted on canvas, first of all he traced a sketch of the whole—very simplified; then he filled in successively the detail of each part of the body, drawing here the design of a head, there an arm or a foot; finally, that which he had obtained by the study of these various pieces, he took as the foundation in a complete composition whose unity has been admired for centuries.

Thus is the New Mechanics made; one, but complex, not capable of being born in one push; a single effort has not sufficed to create it, at one and the same time harmonious in its whole and minutious in it details; in distinguishing the various outlines which have, one after the other, prepared the definitive work, we analyse the composition of this work; we do not break up its unity.

It is not, therefore, in lack of unity that the New Mechanics differs from the Old Mechanics; it differs from it in the complexity of its principles.

The Old Mechanics had thrust to the extreme the simplification of fundamental hypotheses; these hypotheses it had condensed into a single assumption: Every system is reducible to a set of material points and of solid bodies which move in accordance with Lagrange's equations. And even, with Hertz, it had pushed still further on and got rid of real forces in its equations.

The New Mechanics is not, at this point, consumed with simplifying its principles; when it judges it to be necessary it does not hesitate to increase the complexity of its fundamental hypotheses; it admits in its equations terms of various natures and various forms, viscosity terms, frictional terms, terms for hysteresis, electrokinetic energy—whenever the Old Mechanics excludes from its formulae such symbols, contradictory to its single principle.

Now reality is complex, infinitely so; each new perfecting of experimental methods, scrutinising facts more deeply, discovers new complications; the human mind, in its weakness, has, in spite of that, striven towards a simple representation of the external world; it suffices him to place the image opposite the object and to compare them in good faith to establish that this simplicity, so ardently desired, is an elusive chimera, an unrealisable utopia. Willy-nilly, the lessons of experiments oblige him to restore to his system the complexity that he had wished to banish. If, despite everything, he wishes to safeguard the extreme simplicity of the fundamental principle, of the first laws of motion, he has to complicate to excess, by means of hidden motions and unseen masses, the geometrical configuration of the systems to which he intends to apply these laws. To what a heartbreaking degree this complication has had to be brought in order not to renounce the seductive simplicity that mechanical explications promised, and this we know well enough.

The Mechanics founded upon Thermodynamics has in no way imposed upon its hypotheses the exaggerated simplicity that the Old Mechanics required; it has borne whatever they caused, as numerous as varied, whatever they expressed by the most complex formulae. This greater breadth left to the choice of principles is shown to be happy and fertile. To obtain a satisfactory agreement between sensible reality and the mathematical schema that must be substituted for it, it has no longer been necessary to complicate this latter beyond measure; if the beginnings of Mechanics are a little less simple than in the past, the development of physical theories is pursued with an ease unknown until then.

This aptitude for being moulded around facts and to espouse their main features the New Physics has therefore acquired in disembarrassing itself of certain requirements which bridled the Old Mechanics. Amongst these requirements the first and most essential is that which aspires to reduce all the properties of bodies to quantities, shapes and local motions; this requirement the New Physics resolutely repulses; it admits into its arguments the consideration of qualities; it renders to the notion of motion all the generality that Aristotle attributed to it. There is the secret of its marvelous flexibility. By that, in fact, it is freed from the consideration of these hypothetical mechanisms which were repugnant to the natural philosophy of Newton, from the search for masses and hidden motions whose sole object is to explain qualities geometrically; delivered from this labour that Pascal proclaimed to be doubtful, painful and useless, it can, in complete freedom, devote its efforts to more fertile works. Similarly, Alchemy has remained a sterile study as long as it has persisted obstinately in resolving all bodies into salt, sulphur, quick-silver and black earth; from the day upon which Chemistry resigned itself to regarding as simple the substances that it never succeeded in decomposing, it became a science of an admirable fecundity.

The creation of this Mechanics founded upon Thermodynamics is therefore a reaction against atomistic and Cartesian ideas, an idea — quite unexpected by even those who had most contributed to it — to the profoundest of the peripatetic doctrines.

Thus, by a counter-revolution opposed to the Cartesian revolution, the New Mechanics returned to the traditions of Scholastic Physics, so long and so violently decried; but this counter-revolution abandons nothing of the Cartesian conquests. Cartesianism had wished to banish qualities from Physics in order that it might hold forth upon Physics and the language of mathematics; the New Mechanics argues about qualities, but, in order to argue with precision, it accounts for them by numerical symbols; the daughter of Aristotle, in that it is a theory of qualities, it is also a daughter of Descartes, in that it is a Universal Mathematics; in it there begin to converge the two tendencies which, for so long, have drawn Science and Nature in opposite directions.

This trait, furthermore, is in some sort the characteristic of scientific transformations whose phases we have just retraced. Mechanical systems have followed one another in number and variety; but none of them has disappeared without leaving a rich heritage of new ideas to that which supplanted it. Each worker had conceived the plan of an edifice and hewn out his materials for realising this plan; the edifice tumbled down, but the materials which had served for building it appear quite in place in the new monument. Through the vicissitudes which piled up one upon the other against the ephemeral theories, a Guiding Idea seems to be on the lookout to ensure that no sincere effort to find the truth does not remain sterile and in vain. The conscious creator of a mechanical doctrine is the unconscious precursor of the doctrines that will replace it. Let us cite an example: Lagrange only thought to study systems where all is figure and local motion; he intended only to leave the greatest possible indetermination possible in the variable quantities that would represent this figure and this motion; and it it here that without his knowing he chiselled the mould in which the Physics of quality was to be cast, that he wrote the formulae upon which were to depend not only local motion, but, further, the motions of alteration, of generation and of corruption; all that is essential in Lagrange's Statics is found, a hundred years later on, in Gibbs' Chemical Mechanics.

The development of Mechanics is therefore, properly speaking, an *evolution*; each of the stages of this evolution is the natural corollary of the stages that have preceded it; it is the chief part of the stages which will follow it. Meditation upon this law has to be the theoretician's solace. It would be quite presumptuous to imagine that the system for the achievement of which he works will escape the fate common to the systems that have preceded it and will merit lasting longer than them; but without vain boasting, he has the right to believe that his efforts will not be sterile; through the centuries the ideas that he has sown and germinated will continue to increase and to bear their fruit.

SUBJECT INDEX

Absolute temperature 58,132
  Carnot's Principle and — 131-133
  internal potential and — 134-140
action 123
  secondary — 45
actuality, in 3
aether 86
affinity 13
atom 9
  vortex — 89-93

Calorific
  — coefficients 139
  normal — capacity 154
canonical distribution 65
Carnot
  — cycle 131
  —'s Theorem 132
category 2
chemical
  — equilibria
    false — 161-169
Clausius inequality 147
cohesion 13
conduction flux 70
conservation
  — of energy 118-121
  — of kinetic energy 74
  — of vortex tubes 91

constraint 24
  — actions 136
  — conditions 114
  — equations 24
  — forces 38,41
    fictitious — 37
corruption 4

Dead force of a system 34
displacement flux 70
domain
  — of experimental facts 111
  — of instruments and measuring processes 112
  — of theory 111
dynamical systems 42

Electric
  — flux 179
  — motion 179
electrodynamic
  — potential 70,181
  —al induction 180
electrokinetic energy 179
Electromagnetic Theory of Light 183
electrostatic potential 70
endothermic reactions 159
entropy 80
  — of the Universe always increases 81

equilibrium
  displacement from — 158
  stability and — 153-160
  — of frictional systems 166
  — and motion 116-117
  stability of — 36
  states of false — 167
  states of true — 167
  statistical — 50
  system in — 25
exothermic reactions 159
extension 2

Fluid 4
force 23
  dead — 34
  fictitious — 84
  live — 34
free energy 135,152
frictional systems 163

Generalised
  — accelerations 34,69
  — forces 26
  — inertial forces 34
  — velocities 34,69
generation 4
geometric systems 42

Hidden
  — motions 77,78
  — properties 16-21,105
  — qualities 2
hydrostatic pressure 31
hysteresis
  — coefficient 174
  magnetic — 173

Incompressibility 42
independent variables
  — defining a system 26
inertial force 33
inertialess variable 142
internal potential
  — and absolute temperature 134-140
internal thermodynamical potential 135

isentropic
  — modification 151
  — stability 154
isomoric coupling 63
isothermal
  — adiabatic system 148,151
  — displacement 158
  — stability 154

Kinetic
  — energy 35
  — equation
    — and usable energy 150-152
  — Theory of Gases 47-53

Lagrange's
  — Analytical Mechanics 37-46
  — Dynamics 32-36
  — multipliers 29
  — Statics 22-31
Law of Phases 138
live
  — energy 35
  — force
  — equation 35
  — of a system 34
local motion 116
locomotion 4,6

Magnetic displacement current 71
mechanical
  — analogies 53
  — explications 1-103
    general considerations on — 94-103
  — of physical phenomena 1
  — theory
    — of electricity 68-72
    — of heat 54-67
Mechanics
  Analytical —
    Lagrange's — 37-46
  Atomistic — 9-11
  Cartesian — 5-8
  Chemical — 117
  Hertz's — 83-88
  Newtonian — 12-15

INDEX

Peripatetic — 1-4
Physical — 40,117
Poisson's — 37-46
Rational — 23
method
 analytic — 96
 synthetic — 95
model 102
modification
 reversible — 127-128-130,162
 virtual — 111-114-115,116,122
monocyclic system 61
 — in equilibrium 62
motion 3,116
 —s of alteration 116
 natural — 75
 ordered —s 81
 perpetual — 62
  impossibility of — 73-82
 random —s 81
 reversible — 75
mutual energy of two systems 122

Natural state 44
New Statics 117
normal variables 134

Perfect gas 113,133
permanent alterations
 — and hysteresis 170-177
phases 157
physical lines of force 67
Physics of Quality 105-110
place 2
polarisation intensity 179
Postulate
 Clausius — 131
 Kelvin — 131
potential 28
 in — 3
 — energy 35
power 23
Principle
 d'Alembert's — 32-36
 Carnot's —
  — and absolute temperature 131-133
  — of Conservation of Energy 35,118,121

 —s of Electrodynamics 181
 — of Equivalence of Heat and Work 56
 — of General Dynamics 141-145
 Helmholtz's — 154
 Sadi Carnot and Clausius — 58
 — of Virtual Displacements 24
 — of Virtual Velocities 22-31
purely fictitious generalised force 29

Quality 2
 hidden — 2
quantity 2
 work and — of heat 122-126

Real energy 35
remission 2
reversible
 — modification 127-128-130, 162
 — motions 74

Solid 4
stability
 — and displacement from equilibrium 153-160
substance 2
supplementary relations 146-149

Temperature scale 49
Theory of Migration of Energy 95
thermodynamics 59
thinkers
 abstract — 99
 imaginative — 99
Total Energy of the Universe is constant 81
transformation value 58,80

Uncompensated transformation 80

usable energy 151
useful mechanical effect 152

Virtual
 — displacements 25
 — modification 111-114-115, 116,122
 — work 25,123
vanishing — 25

void 9
vortex
 — atom 89-93
 — line 90

Work 25
 — by inertial forces 124
 — and quantity of heat 122-126